EVOLUTION AND EXPRESSION OF
HORMONES IN FISH

EVOLUTION AND EXPRESSION OF

HORMONES IN FISH

Dipti Nagar

Oxford Book Company

Jaipur ▪ New Delhi

Published By

Oxford Book Company

Regd. Off. "Bony Residency", Gate No. 2
Opp. Tilak Public School, Vishweswariya Nagar,
Gopalpura Road, Jaipur - 302018, India
Ph: 0141 - 2761280, Telefax: 2761381
email - oxfordbookjpr@gmail.com
website - oxfordbookcompany.com

Delhi Office: 102, 1st Floor, Satyam House,
4327/3, Ansari Road, Darya Ganj,
New Delhi - 110002, India
Ph: 011-45652440

Distributed By
KEYA INTERNATIONAL PRESS LLP
E-mail: keyainternationalpressllp@gmail.com

Published in India, 2025

ISBN : 9789391096403

Typeset by Shivangi Computers, Jaipur

Printed and bound in India by Thomson Press Ltd.,
New Delhi

Contents

Preface

Fish make up over half the total number of living vertebrate species and exhibit a level of variation in sexual behaviour that is unrivalled among other vertebrates. The early organization and adult reorganization of neurobiology and behaviour generate sexual plasticity in fishes, and these processes can be socially controlled. The major brain areas and neuroendocrine factors that control the development and adult maintenance of reproductive behaviour and physiology in fish do not differ substantially from those in mammals, suggesting that the examination of fish provides useful insights into the generation of sexual variation in all vertebrates. The explosive radiation of the percomorph, or advanced, fishes resulted in a tremendous diversity of life histories that has been generated by the invasion of new habitats and characterized by both genotypic adaptation and phenotypic polymorphism. Adaptation to new habitats generates a host of changes in the behaviour, anatomy, and physiology of these animals. As a result, percomorph fishes provide a host of natural experiments to investigate the responses of hormones, brain, and behaviour to novel environmental challenges.

Among ûshes, the teleosts include the most commonly recognized fish species (e.g., bass, perch, and flatfish) and represent the most diverse of all vertebrate taxa. In the teleosts, there are at least six classes of reproductive phenotypes : (1) reproductive dysfunction, including permanent infertility and environmentally induced (e.g., social or xenobiotic inhibition) reproductive compromises that result in deficits that vary in both duration and intensity; (2) a single phenotype in a sex, the basic pattern common to most ûshes and vertebrates (gonochorism); (3) a single bisexual form that functions as a simultaneous hermaphrodite; (4) multiple phenotypes in a sex that often take the form of two distinct male phenotypes that use different behavioural approaches to acquire mates (known as alternative reproductive tactics); (5) singular reproductive transformations resulting from unidirectional adult sex change; and (6) serial reproductive transformations characterized by back-and-forth (or serial) sex change.

This review concentrates on the last three categories. Sexual polymorphism in any given population represents a dynamic interaction between the genetic potential of a species and the environmental–social modulation of this genetic background. For this reason, a holistic approach that examines the system at multiple levels of analysis may be most effective. For example, many fish lay benthic eggs, and this requires one parent to protect the eggs at least until they hatch. Most species exhibit male parental care—the female leaves once the eggs are attached to the bottom and the male is left to care for them. Males attain reproductive success by holding high-quality territories, constituting a mating system known as male resource defence polygyny. A large male that can defend a high-quality territory attracts many females, whereas most males attract very few. This skew in reproductive success can drive the development or evolution of reproductive plasticity (e.g., small sneaker males or sex-role reversal). The generation of this degree of sexual plasticity in turn requires the modification of a host of behavioural, physiological, endocrinological, and anatomical characteristics. Hence, an important tool in understanding sexual plasticity is delineating the characteristics that are useful for defining sexual phenotype. Even among gonochoristic species, a large number of behavioural, somatic, and gametic characteristics vary in and between the sexes.

The book will not only meet the requirement of students but will also be useful as a guideline to the teachers and researchers.

— Dipti Nagar

Chapter 1

Hormonal Manipulation of Fishes

Aquaculture is the controlled reproduction and growing of aquatic organisms for food or ornament. Current trends in human population growth, habitat and natural stocks destruction versus advances in world agricultural and fisheries production clearly show the basic need for developing and implementing aquacultural technology.

Widespread use of hypophysation (use of the pituitary gland, a part of the brain and endocrine system of humans and fishes) has permitted breeding of a number of species otherwise not spawned in confinement (e.g. kuhli loaches, Acanthopthalmus kuhlii, dolphinfish, Coryphaena hippurus). Many of the important cultured pet and food fishes fall into this category. Previously fingerling-stock had to be collected from the wild. This supply was seasonal and unpredictable. By manipulating hormonal and environmental factors seed can be produced all year long; reducing cost, disease, and incidence of trash fish (undesireable contaminants) and predators.

Hypophysation involves the injection of gonadotropins (hormones=chemical messengers produced by the pituitary, that stimulate the reproductive organs) from one animal into another. Among vertebrates (fishes, amphibians, reptiles, birds and mammals) these are not very species-specific. Sometimes it has been found better to mix mammalian pituitary extracts and steroids in varying concentrations with fish gonadotropin. There is increased success in the use of prepared mammalian hormones in breeding procedures, preservation of fish sperm, maintenance of brood stock; yielding reliable sources of fish seed and acceleration of breeding research.

Problems relating to standardization of dosage, determination of maturity of recipient fish and nutritional requirements particularly need to be further explored. Current knowledge on the use of pituitary hormones in fish culture and technology regarding collecting, processing & storing of glands & methods of application have been covered comprehensively in the literature. The first experiments regarding reproductive behavior (in female mice) and replacement therapy using pituitary implants. The first use in fishes was in 1930 by B.A. Houssay of Argentina. He induced premature birth in a viviparous (livebearing, like guppies) fish by injecting pituitary glands recently removed from other fish. Other South Americans, like von Ihering, a Brazilian, used this technique for aquaculture extensively beginning in 1934.

Soviets attempted to induce captive sturgeons to spawn in 1932 by means of mammalian hormones. They were unsuccessful until 1937, when N.L. Gerbil'skii produced eggs and sperm in a number of the sturgeon, Acipenser stellatus that had been intracranially injected with 1-2 fresh pituitaries of conspecifics (the same species). As of 1957 the Soviets have obtained all their sturgeon eggs for culture from pituitary treated fishes, very important to their caviar industry. As of 1963 the Chinese have been able to supply all the seed used in carp culture (big-head, silver, common & other food species) by means of hypophysation.

Methods of Induced Breeding

Fish are induced to spawn by altering environmental parameters, particularly temperature and photoperiod, and hormonal constitution. These variables are generally constant for the species but vary widely among cultured food fishes.

Environmental Factors and Diet

Environment and diet are the principal sources of phenomena determining hormonal secretion. Several factors, photoperiod, temperature, metabolites, pheromones, light strength and temperature shock control release of gonadotrophins, which in turn induce gamete (eggs & sperm) production and concomitant reproductive behavior. Reproduction may occur out of season by manipulation of temperature and photoperiod (how long the light is on daily) alone.

An example of improved culture just through manipulation of environmental factors may illustrate the point. The Ayu, *Plecoglossus altivelus*, a smelt-like fish cultured in Japan during the summer months, normally spawns in October-November., after which most of the fish die. Due to lower water temperatures that time of year, it's difficult to maintain live food organisms in adequate densities in rearing ponds to feed fry. Adjustment of photoperiod resulted in successful acceleration of sexual maturation (August), enabling rearing of fry when natural food was abundant. Spawning was also delayed to prolong the life of adults. The light period was extended to 18 hours per day with artificial light (August through October). Reproduction was retarded & adults were marketed out of season in February.

Temperature

The relative importance of temperature varies per species, investigator and experiment, but all agree a certain range and lowering or elevating optimizes results. Smiglieski (1975) stated that temperature was the controlling factor in his work with flounders. Yamamoto et. al (1966) observed spermatogenesis is goldfish, Carassius auratus below 14 degrees C & accelerated spermiation above 20 degrees C. Often, if much temperature fluctuation occurs, reproductive behavior will cease. Haydock (1971) has observed temperature threshold below which gulf croaker will not hydrate or ovulate. For each species or sometimes race (a sub-specific classification) there are limits, both high and low, which eliminate reproduction physiologically and behaviorally.

Light

As with temperature, there are optimal amounts of strength, quality and duration of light in relation to reproductive behavior. On work with bluegill sunfishes it has been found that a longer photoperiod 16light/8dark versus 8L/16D induced a better gonosomatic index (G.S.I.= ovary weight divided by total body weight times 100). Males were more aggressive, dug more nests and fertilized more eggs.

Hoar (1965b) sums up the effect of longer light periods by stating that they stimulate secretory activity of anterior pituitary, inducing pre-sexual behavior. The pituitary LH, Lutenizing Hormone (one of the gonadotropic hormones produced in the brain) activates interstitial tissue of the gonads which produce gonadal steroids that in turn dominate sexual phases, taking complete control during prenatal phases. Several investigators have shown that elevated temperature and prolonged photoperiod increased gonad maturation in the green sunfish, Lepomis cyanellus.

Other

Photoperiod and temperature have received most attention & are generally considered to be of greatest importance in inducement of sexual maturity and spawning. Other factors such as the effects of other environmental stimuli; meteorological (rain, floods, etc.) & water conditions (pH, Ammonia, $C0_2$, turbidity) on controlled breeding of food and ornamental fishes has been investigated. In several species, the presence of con- (the same species) and/or hetero- (other) specifics is important. In bluegills and some cichlids it has been found that when males are present, the most aggressive female has greaater ovarian development. Another application of environmental control is employed in mullet culture in Israel. By keeping the adults in freshwater ponds and not allowing access to the sea where they migrate to spawn, fish-culturists are able to extend the spawning season.

Diet

Researchers have demonstrated gonadal regression induced by restricted diets accompanied by an inversion (opposite) of the lasophil/ acidophil (cell types distinquished by their staining characteristics) ratio of the mesoadenohypophyis (an anterior part of the pituitary gland in ours and fishes brains) and a reduction in gonadotrophin content. These conditions were reversed with adequate food. Some authors have stressed the interaction between photoperiod and diet and pointed out that optimum benefit from photoperiod adjustment can be over-rided by poor diets. Hmmm.

C. Hormonal Manipulation

Criteria: Biological and Chemical Assay Several methods are applied towards determining sex and sexual maturity of spawners. Externally there may be sexual dimorphism (structural differences between the sexes) in the species and/or differences that occur during sexual ripeness. Such is the case with the channel catfish, Ictalurus punctatus. Females assume a dropsical appearance with raised and reddened genitalia. The males develop swollen head musculature and well defined genital papilla.

A practical and precise way to measure oocyte (egg cell) development is through inserting a sterile polyethylene or glass catheter in the oviduct and removing egg samples. The eggs are cleared in successive concentrations of ethyl (grain, drinking) alcohol (30, 50, 70, 90, 100%) then xylene (an organic solvent) for 5 minutes each. Eggs are removed by catheter before each hormone injection to determine gonadal development.

Physiological parameters are also valuable as sources for assays. Elevated plasma (the fluid portion of blood, the same as for humans) calcium levels have been reported during gonad development in many species. Peterson & Shehadeh (1971) reported a four time increase in calcium and two time plasma copper in striped mullet, Mugil cephalus, under methyl testosterone treatment.

For examination, gonads are removed from the body cavity, weighed to 0.01 gram, labelled, wrapped in cheese cloth, fixed in 70% ethanol (ethyl alcohol) for 30 minutes and 100% for one hour. Xylene is used for a clearing agent. Infiltration and final embedding is done in fresh paraffin (wax). Typically microscope slides are produced by slicing these specimens on a tool called a microtome at 8 micron thickness. These tissue slides are stained with a compound called Delafields iron hematoxylin and counter stained (for contrast) with eosin. The specimen is permanently covered with a thin glass coverslip and permount.

When you examine these prepared sections under a microscope you can judge the sexual readiness oth your stock. In sexually immature males the lumen (opening) of the seminiferous tubules is unformed. Later primary and secondary spermatocytes develop in the lumen. Still later spermatids and spermatozoa (these are all developmental phases of male reproductive cells) are seen free in the lumen. After spawning the tubules are completely occluded (blocked) and testes become highly vascularized (serviced with blood vessels). In immature females only the oogonia & primary (previtellogenic) oocytes are present. At maaturity secondary oocytes distributed by oolemma and centripedal yolk precursor form in the cytoplasm (the fluid portion of cells). The pituitary may be examined in sample individuals. They are dissected and promptly fixed in Bouin's or Zenker's for 24 hours. They are subsequently passed through alcohols and stored in 70% ethanol, further processed and sectioned. In the channel catfish the adult pituitary double from one to two millimetre maximum length at sexual ripeness. The mesoadonhypophysis contains gonadotropic and somatotropic cells; the former enlarge in ripe specimens. In this way the mesoadenohypophysis becomes large during sexual maturity.

Problems of Standard Dosage

The endocrine (=cry internally) system acts like a chemical link between an organism and it's environment. Fluctuations in the external environment and diet mediated through the central nervous system (CNS) affect pituitary secretion. As one of several target tissues the gonads are effected. The CNS acts quickly, temporarily and specifically. The endocrine works in concert and antagonistically with the CNS but slower, longer-lasting and more generally.

It is well established that in all vertebrates the pituitary gonadotropins and gonadal steroids are the hormones most directly involved in the regulation of reproductive behavior. Pituitary hormones may act by stimulating secretion of gonadotropic hormones, which in turn regulate behavior. Also, individually these hormones may influence behavior indirectly by their role in the development of secondary sexual characteristics.

In turn the CNS-pituitary-gonad axis is influenced by internal and external factors acting through the sense organs and brain. Studies on birds have emphasized that coordination and instigation of components of reproductive behavior are dependent on a continuous interplay of external and internal factors; hormones effect behavior and behavior effects the endocrine system.

There is some evidence that hormones other than those of the anterior pituitary and gonads play a major role in reproduction. Neurophysiological and thyroid hormones most outstandingly, but other hormones affect reproduction indirectly by their effects on general metabolism and growth. All of the endocrine glands are either directly of indirectly involved, since reproduction requires profound metabolic adjustments, particularly in females.

Injections: Two methods of injection are in wide practice. Intramuscular, in the flank just below the dorsal fin and behind the gill cover. This method is safer but slower working. Interperitoneal injections are made into the body cavity. They are faster acting but involve a greater chance of injury or death.

Dosage, latency period and frequency of injection have not been well determined for any one species; investigators adjust variables according to specific circumstances. This has yielded divergent results, duplication of effort and failure. Problems of determining exact dosages lie in two factors. 1) the largest fish does'nt necessarily need the largest dose; this is more dependent on the hormonal development of the individual. 2) Exact dose of any preparation depends on determination of it's gonadotropic content. Towards the ends of standardization an international standard animal for bioassay has yet to be agreed upon. The Soviets use the Vy'un unit (U.V.) based on the loach, Misgurnis fossilis, & the frog unit (Curimata sp.). The Brazilians use the fish unit (Fontenele 1955). A Wagur unit using Clarias batrachus (the walking catfish!) was advocated by Das & Khan (1962). In the U.S.A. changes in the gonadosomatic index (GSI) of hypophysectomized (with pituitary gland removed) cyprinodontid (killifish family) Fundulus heteroclitus have been often used. More recently, a bioassay assays using seminal fluid viscosity have been used. Goldfish spermiation was used by Yamazaki & Donaldson (1968) in response to purified salmon gonadotropin. Otsuka (1956) monitored uterine and ovarian changes in immature mice. The Galli-mainini reaction and spermiation or hydration in goldfish has been used most frequently. For convenience the wet weight of pituitary or dry weight (acetone dried) of pituitary powder/weight of fish is unfortunately still used versus chemical analysis of hormones to determine which are present, active and their synergistic-antagonistic effect on each other.

So it can be seen that due to the variety of test animals as standards, & lack of test for definitive presence and concentration of hormones, comparison of results is difficult, and therefore the data are mostly of local benefit.

More precise bio-physico-chemico-assay methods such as incorporation of radio-phosphorus by gonads is in it's infant stages. (This is a specific labelling and measuring technique)

Maturity of Recipient Fish

Variation in a species in terms of gonad development is intensified in confinement. Variability of results is due to individual variation in sex or season, especially considering crude bioassay techniques. Effectiveness of hypophysation (injection with pituitary hormones) is dependent on the stage of reproductive development of recipients. Aquaculturists recommend that when applying other workers' data; that matching the dosage to the physiological state of the recipient can best be done by administering doses in graded series from above or below those recommended by other investigators. Almost always slightly greater success is achieved with the most intense regime (injecting less hormone more frequently).

If the species has not been spawned before they suggest continuous injections, every 24 hours, with a wide range of doses until spawning does occur. As has been mentioned, care must be exercised in assaying sexual readiness in spawners. Parameters generally adopted have frequently proven unreliable. An example of this is females with enlarged abdomens, reddish colouration & protrusion of the cloacal region may be due to engorgement of the intestine, or disease, even during the spawning season.

Variability in Pituitary Gonadotropin Related to Phylogeny, Sex & Season

In order to determine dosage, effect and success of treatment it is vital that fluctuations in gonadotropic activity related to phylogenetic specificity (basically how closely species are related to each other evolutionarily), sex and season be known for each species. Much controversy exists as to specificity in gonadotropin as it relates to vertebrate phylogeny. Work with homoplastic (derived from the same species) pituitaries has produced best results. Little correlation exists between phylogenetic affinity and effect of pituitary gonadotropins. Where distantly related donor-recipients are used the general practice has been to increase dosage.

The early literature on sex related differences in the gonadotropic activity of fish pituitaries has been outlined by Pickford & Atz (1957). Much of it and later reports are contradictory. It is more or less agreed that sex-related differences in pituitary extracts are usually minimised by the usual preparation of mixed extracts of both sexes in excess of threshold requirements.

Seasonal fluctuations in gonadotrophic activity are not uncommon in almost all fish species examined. Pituitaries collected for hypophysation are almost always taken from mature fishes during spawning season to be used immediately or dried and prepared for later use. (Kind of hard on your breeders, eh?)

Use of Mammalian Hormones

Advantages of using mammalian hormones include:

(1) Reliable source, available all the time.

(2) Easily stored in lab and retain biological activity for a number of years.

(3) More uniform than crude pituitary preparations, permitting controlled experiments and repeatable results.

(4) Cost is about the same.

(5) Saves valuable gravid fish, especially where homoplastic pituitaries are used.

(6) Does away with the tedious task of collecting, preserving, processing & distributing pituitary material.

There is not much in the literature on the use of such hormones for practical breeding of ornamental fishes; quite a bit on food species. Several Indian workers have worked extensively with Heteropneustes and mammalian hypophysial hormones, placental gonadotrophins, gonadal hormones and corticosteroids finding that human chorionic gonadotropin (HCG) and some of the corticoids, especially deoxycorticosteroids could be used profitably to spawn food fishes as a substitute for pituitary extracts. Chorionic gonadotrophin was also found to be an effective ovulating agent in Clarias batrachus. Snead

& Clemens (1959) established methods for spawning channel catfish, goldfish, and white crappie, Pomoxis annularis with injections of chorionic gonadotrophins. At the price of $5.65 per 10,000 international units (IU) of APL (Anterior Pituitary Like) they established a cost of 45 cents per pound of fish spawned, comparable to established costs of fresh pituitary preparations. Other reviews of successful use of mammalian hormones exist detailing one's effectiveness over another, acting singly and together.

Purified fish gonadotropins are available to the fish breeder, especially fractionated salmon pituitary. Much work has been done with these and food fishes; not much reported for pet fish.

Pituitary Extracts, Use and Preservation

In a fish, the pituitary is concerned with regulation of growth, pigmentation, adrenal cortex, thyroid gland and others, as well as reproduction. Cyclical changes in the pituitary are striking, showing increase and decrease in cyanophil cells involved with production of gonad stimulating hormones. The synergic action of growth hormone, thyrotropic hormone (stimulating the thyroid gland) & sex steroids is indicated by the better results obtained with crude rather than purified gonadotropic hormones.

The pituitary regulates both gametogenesis and steroidogenesis in fishes. Two seperate gonadotropins Follicle-stimulating hormone (FSH) and Lutenizing hormone (LH) are physiologically distinct in the tetrapods (= "four legs", the rest of the vertebrates beyond the fishes, i.e. the amphibians, reptiles, birds & mammals). There is evidence of a single protein in fishes. When fish pituitary extracts are tested in tetrapods, both FSH and LH effects are elicited.

With rare exceptions induced spawning by hypophysation has been restricted to females. This is often due to the availability of male seed. Collection of pituitaries involves chopping the head off the fish above the eyes, lifting the brain and removing the gland from the base of the cranium.

Preserving the gland is done by freezing immediately with dry ice or dehydrating in acetone. This is done by changing the acetone several times every 12 hours. This is then stored in a desiccator (dryer). It lasts up to two years but the effect is not as good as fresh.

When preparing, care must be taken to not introduce foreign proteins. Combined with pituitary this sometimes causes very strong rejection. Therefore purification is necessary. Whole fish pituitaries may be ground up or injected whole. If ground, they are suspended in physiological saline or distilled water.

Gonadal Steroids

In all fishes investigated development of secondary sexual characteristics (are they growing facial hair?) has been shown to depend on gonadal steroids. Several androgens (male sexual hormones like testosterone) and estrogens (the female equivalent) have been identified from the tissues and blood of fishes. Evidence shows that some fish gonadal steroids have the same biochemical pathways as higher vertebrates. Among the steroids found are testosterone, androstenedione, 17 beta estradiol, androsterone, 11 keto testosterone, conjugated testosterone and progesterone. As with higher vertebrates bot males and females have compounds in common but in varying concentrations. Hoar (1962b) presents evidence that reproductive behavior becomes dependent (in a feed-back loop with the pituitary) on gonadal hormones and less sensitive to pituitary gonadotropic hormones. Acquistion of nuptial colouration & onset of breeding behavior has been correlated with increasing steroidogenesis.

Even in a paper of this scope it's possible to draw general conclusions about the role of hormones in concert with environmental factors in control of spawning behavior of fishes. In male fishes of several species, secondary sex-morphological changes are dependent on gonadal hormones. In several species, components of reproductive behavior are under gonadal control, yet in other species much of all reproductive behavior persists after castration. (Is this evidence of learned behavior, conditioned response?)

Ovariectomy abolished sexual response in several species examined. In the guppy sexual behavior reappears some time after ovariectomy. There has been no unequivocal demonstation that administration of estrogen or progesterone will induce reproductive behavior in fishes. Some of the reasons allowing for the differential success in use of hormones to control reproduction in fishes are the phylogenetic differences in protein and peptide hormones of the pituitary. On the other hand, the similarity of steroids suggests there is little problem in phylogenetic specificity in applying steroid therapy.

I'd like to see controlled studies on the use of steroids on colour and growth augmentation in sport mutations of oscars, Astronotus, guppies, and discus, Symphysodon among other hormone-treated species from the far east, in terms of related reproductive viability, longevity and vitality. How about you?

Pet Fish Conclusions:

1) The effects of temperature, light (duration, quality & strength), metabolites ("fish wastes"), are very important , and many other factors to a lesser extent, in getting stock up to and increasing success in reproduction.
2) There is an extensive body of knowledge in scientific literature on the behavior and physiology of fishes, in particular food & game species.
3) That you probably "knew" much of this material intuitively from your association with aquarium species, and indeed could write a "scientific" paper of substantial importance concerning your experience, if not a thesis.

Materials and Methods

Nomenclature of GnRH Ligands and Receptors: GnRH1 denotes the peptide that regulates reproduction in vertebrates, and we use mGnRH1 to distinguish the mammalian form. For other GnRH peptide sequences, we show amino acid changes relative to the mGnRH1 sequence (35). GnRH peptides used in this study were mGnRH1 (pGlu-His-Trp-Ser-Tyr-Gly-Leu-Arg-Pro-Gly-NH_2) and the three forms of GnRH found in *A. burtoni*, [Ser^8] GnRH1 (also called seabream GnRH), GnRH2 ([His^5,Trp^7,Tyr^8] GnRH) (36), and GnRH3 ([Trp^7,Leu^8] GnRH, also called salmon GnRH) (35, 37). GnRH-Rs are designated as in GenBank with the addition of a three-letter superscript, which is the conserved motif in EC3 (SHS, PEY, PPS, or SDP) to indicate their GnRH-R type classification. Hence, the receptor identified by our experiments becomes GnRH-$R1^{SHS}$, because it was designated GnRH-R1 in GenBank and contains the SHS sequence in EC3. Correspondingly, the other *A. burtoni* GnRH-R is named GnRH $R2^{PEY}$. In receptors where the EC3 motif is not identical to the conserved motif, the motif in the receptor is used (*e.g.* GnRH-$R1^{PDY}$ for one of the catfish receptors, which has PDY instead of PEY in EC3).

Animals

A. (Haplochromis) burtoni derived from a wild-caught population were raised in aquaria under conditions similar to their native equatorial habitat in Lake Tanganyika in Africa (pH 8, 28 C) (28). Both male and female fish (3–4 cm length) were used in these experiments. To collect brain tissue, fish were killed via rapid cervical transection. All work was performed in compliance with Stanford University (APLAC) guidelines.

Phylogenetic Analysis of GnRH-R

We used phylogenetic analysis to situate *A. burtoni* GnRH-Rs with respect to previously cloned GnRH-Rs. All publicly available (February 14, 2007) vertebrate full-length GnRH-Rs were obtained from GenBank. Multiple GnRH-R sequences were aligned using the predicted sequences of GnRH-R polypeptides (ClustalW program in MEGA3.1) (39). This multiple sequence alignment was used to generate a neighbour-joining tree with bootstrap values using the Jones-Taylor-Thornton matrix. Lamprey GnRH-R was used to root the phylogenetic tree. Full species names and GenBank accession numbers for the receptor cDNAs are listed in the supplemental material.

Functional Assay of GnRH-R1^{SHS}

For functional tests, the *A. burtoni* GnRH-R1^{SHS} cDNA expression plasmid was transfected into COS-1 cells using diethylaminoethyl dextran, as previously described (40). The ligand structure-function relationships of the putative GnRH-R1^{SHS} were directly compared with *A. burtoni* GnRH-R2^{PEY} (32) by measuring ligand-stimulated inositol phosphate (IP) production, as described previously (40), except that IP was extracted with formic acid (10 mm). Ligand concentrations ranged from $10^{"10}$ to $10^{"5}$ m. Data points were determined in duplicate, and EC_{50} values are the means of three separate experiments.

For competition binding assays, [His^5,d-Tyr^6] GnRH1 was radioiodinated as previously described (41). Transfected cells were washed twice with cold PBS and incubated (5 h at 4 C) with ^{125}I-labelled [His^5,d-Tyr^6] GnRH1 (10^5 cpm) in the absence (B0) or presence of various concentrations ($10^{"9}$ to $10^{"5}$ m) of unlabelled peptides in 0.5 ml HEPES-buffered DMEM with BSA (0.1%). Unbound ^{125}I-labelled [His^5,d-Tyr^6]-GnRH was removed by washing twice with PBS, and cell-associated radioactivity was solubilized in 1 ml 1 m NaOH and counted. Nonspecific binding was determined in the presence of excess GnRH2 ($10^{"5}$ m). Data points were determined in duplicate, and the IC_{50} values are the means of three separate experiments.

Localisation of GnRH-R1^{SHS} mRNA Transcripts

GnRH-R1^{SHS} expression was localised using PCR on tissue samples from adult *A. burtoni* (brain, pituitary gland, retina, gill, heart, intestine, kidney, liver, muscle, ovary, retina, spleen, and testes) and *in situ* hybridization using antisense receptor-specific probes in the brain and pituitary.

In Situ Hybridization

In situ hybridization methods are described in detail elsewhere (33), so only essential details are given here. Reproductively active males (n = 3) were killed by rapid cervical transection, and brains were removed and placed in sterile PBS. Tissue was frozen in mounting medium (OCT; Tissue-Tek, Torrance, CA) on dry ice, sectioned coronally, and stored at “80 C until use. Probes for GnRH-R1^{SHS} and GnRH-R2^{PEY} were synthesized, based on GenBank sequences (GnRH-R1^{SHS}, GenBank AY705931; GnRH-R2^{PEY}, AY028476), and plasmids were linearized to generate both sense and antisense templates for each receptor. GnRH-R1^{SHS} and GnRH-R2^{PEY} probes were from 151-1035 bp and from 55–1129 bp, respectively, and were generated by transcribing with SP6 or T7 polymerase in the presence of [^{35}S]UTP (Maxiscript kit; Ambion, Austin, TX). Slides with brain tissue were treated as described previously (33). Photomicrographs were captured digitally and optimized for clarity.

PCR on Tissue Samples

Tissue from T males (<“6 cm; ovaries from sexually mature female <“6 cm standard length) was homogenized and RNA was extracted (RNeasy Micro Kit; QIAGEN Inc., Valencia, CA). cDNA was synthesized from total RNA from each tissue sample. Sense and antisense primers specific to *A. burtoni* GnRH-R1^{SHS} generated a 420-bp PCR product, whereas sense and antisense primers specific to the *A. burtoni* GnRH-R2^{PEY} generated a 961-bp PCR product from the cDNA template. The GnRH-R2^{PEY} primer pair spans two introns predicted from the conserved GnRH-R genomic structure in fish (12). Positive controls for primers were performed using brain cDNA because *in situ* hybridization assays have shown both GnRH-R1^{SHS} and GnRH-R2^{PEY} expression in the brain (33). Negative controls were performed using the same procedure as for the experimental group without adding cDNA from any tissue. PCR was performed with a 68–60 C touchdown protocol as follows: 3 min denaturation at 95 C, followed by 16 cycles of 30 sec denaturation at 95 C, 30 sec annealing (68–60 C), and 15 min extension at 72 C. These reactions yielded a single product, as revealed by gel electrophoresis.

Immunocytochemical Staining

To localise the somatotropes and gonadotropes, reproductively active fish (T males; n = 2) were killed by rapid cervical transection, and their brains removed and placed in 4% paraformaldehyde in PBS (pH 7.4) overnight and then transferred into a 30% sucrose solution overnight. Fixed tissue was frozen in OCT on dry ice and stored at “80 C. Three series of sections were cut on a cryostat (Microm; Zeiss, Oberkochen, Germany) in coronal or sagittal planes at 14 μm.

Sections were rehydrated (PBS) and incubated in blocking solution (0.3% Triton X-100, 0.2% BSA, 10% normal goat serum diluted in PBS). Slides were incubated separately in three primary antisera [rabbit anti-GH antibody (lot 8502), which labels somatotropes, and rabbit anti-LH antibody (lot 8506) and rabbit anti-FSH antibody (lot 8707), which label gonadotropes in *A. burtoni*, all generously provided by Dr. A. Takahashi] diluted 1:1000 in blocking solution at 4 C overnight. The other set of slides was incubated in blocking solution instead of primary antiserum for the control group. All of the sections were washed in PBS, and the signal was amplified (ABC kit; Vector Laboratories, Burlingame, CA). Sections were incubated in 32 ,32 -diaminobenzidine (Sigma-Aldrich, St. Louis, MO) to visualize the bound antibody and counterstained with cresyl violet to visualize cell nuclei. The slides were dehydrated in an ethanol/xylene series and coverslipped with Permount medium for microscope viewing.

Regulation of GnRH-R mRNA Abundance

Tissue Preparation.

To discover whether expression of GnRH-R mRNA was regulated, either as a function of diurnal time or with respect to GnRH ligand expression patterns, animals (n = 24) were housed in separate aquaria with three animals of mixed sex including one T male in each of eight tanks to provide minimal disturbance of other animals in the room when the animals were killed. Tissue was collected from animals in a single tank at 3-h intervals over a 24-h period beginning at 1000 h (fish were maintained on a 12-h light, 12-h dark cycle with dark-to-light transitions at 0900 h). At night, the aquarium room was accessed through a light-tight door, and all tissue-processing steps were performed in complete darkness using minimal infrared illumination. Fish were killed by rapid cervical transection, and their brains and pituitaries were rapidly removed. Brain and pituitary samples were stored at “80 C until use.

Tissue taken at each time point was separately homogenized in chilled TRIzol (Invitrogen) with a Tissue Tearor and kept frozen in TRIzol reagent at “80 C until RNA extraction.

RNA Extraction and PCR Sample Preparation.

Total RNA was extracted from samples following a standard protocol (TRIzol; Invitrogen). RNA was DNase treated to remove genomic DNA contamination (TURBO DNA-free; Ambion), and 1.0 µg total RNA from each tissue sample was reverse transcribed to cDNA (SuperScript II RNase H reverse transcriptase; Invitrogen).

Quantitative Real-time PCR of A. Burtoni GnRH Ligands and Receptors

Primers for quantitative real-time PCR for [Ser^8] GnRH1 were 52 -cagacacactgggcaatatg-32 and 52 -ggccacactcgcaaga-32 , generating a 128-bp product; for GnRH2, primers were 52 -tggactcctttggcacatcagaga-32 and 52 -ctctggctaaggcatccagaagaa-32 , generating a 126-bp product; for GnRH3, 52 -atggatggctaccaggtggaaaga-32 and 52 -tggatttgggcatttgcctcatcg-32, generating a 11- bp product; for GnRH-R1^{SHS}, 52 -tcagtacagcggcgaaag-32 and 52 -gcatctacgggcatcacgat-32 , generating a 187-bp product; and for GnRH-R2^{PEY}, 52 -ggctgctcagttccgagtt-32 and 52 -cgcatcaccaccataccact-32 , generating a 220-bp product. Three housekeeping genes, actin, 18S rRNA, and tubulin, cloned in *A. burtoni* were used to control for sample differences in total cDNA content. PCR were performed (iCycler; Bio-Rad, Hercules, CA), and reaction progress in 30-µl reaction volumes was monitored by fluorescence detection at 490 nm during each annealing step. Reactions contained 2× IQ SYBR Green SuperMix (Bio-Rad), 10 µm of each primer, and 1 ng cDNA (RNA equivalent). Reaction conditions were 1 min at 95 C and then 35 cycles of 30 sec at 95 C, 30 sec at 60 C, and 30 sec at 72 C, followed by a melting curve analysis over the temperature range from 95 C to 4 C. All reactions were run in duplicate.

Quantitative Real-time PCR Data Analysis.

Fluorescence readings for each sample were baseline subtracted, and suitable fluorescence thresholds were automatically determined by the MyiQ software. To determine the number of cycles needed to reach threshold, the original fluorescence reading data were analyzed using a special PCR algorithm (43). To calculate the relative mRNA amount, target gene levels were normalised using the geometric mean of housekeeping gene mRNA values.

Statistical Analyses

. To test the hypothesis that GnRH ligands and/or receptors were expressed with a diurnal rhythm, we analyzed the mRNA samples taken around the clock for rhythmic variation using generalised additive models (44, 45). This technique identifies nonlinear effects that might produce rhythmicity by analyzing the residuals of a scatterplot smoother fit to the data with a back-fitting algorithm. For the generalised additive models and the linear regressions, we took $P < 0.05$ as the initial significance level and applied a Bonferroni step-down correction to account for family-wise error due to multiple testing. These analyses were computed using R (R:Statistical computing environment; R Foundation, Vienna, Austria).

To identify potential linear relationships between ligands/receptors in a larger data set, we used robust regression (Siegel repeated means) (46) that allowed for the observed nonconstant variance (heteroskedasticity) in the data without requiring data transformation. As a control, we similarly compared the relationships between the ligands/receptors and the Clock 1a gene (GenBank DQ923857). To look for independent sources underlying the data distribution, we used independent component analysis (ICA), which is a recently developed technique related to latent variable and factor analyses. Unlike more traditional approaches, ICA yields unique (except for the sign) solutions to the problem of finding underlying sources of variation by maximising the information contained in each independent source of interaction (44). The resulting sources may thus be interpreted as latent variables containing information about independent sources of variation in the data. We examined the relationships of each of the ligands/receptors with these latent variables by regressing each of the ligands/receptors on each of the independent sources, again using robust regression. We used a consensus source matrix from 100 solutions arrived at using the maximum-likelihood ICA package (computed using R). It is important to note that because the ICA signs are arbitrary, we can discover whether ligands/receptors relate to sources in the same or opposite directions or not at all (*i.e.* the signs can be reversed, but not the relative relationships between variables).

Results

Cloning of Full-Length GnRH-R1^{SHS} from A. Burtoni: Analysis of the full-length putative GnRH-R cDNA sequence from *A. burtoni* tissue shows it to be a second GnRH-R, with conserved Asp residues in transmembrane segments 2 and 7 and the carboxyl-terminal tail that are characteristic of GnRH-Rs that are not mammalian SDP-type GnRH-Rs (47, 48). There are conserved motifs SCAFVT in transmembrane segment 3 and VSHS in EC3 characteristic of teleost receptors as identified by Lethimonier *et al.* (12). This sequence is most similar to other SHS-type GnRH-Rs especially from teleost fish (*e.g.* GnRH-R1^{SHS} *O. niloticus*, GenBank AY705931). Consistent with our proposed nomenclature, we named this receptor GnRH-R1^{SHS} (12,

33, 34). The sequence also has residues conserved among all GnRH-Rs, Asp^{107}, Trp^{110}, Asn^{111}, Lys^{130}, Asn^{217}, and Tyr^{298}, which are equivalent to residues Asp^{98}, Trp^{101}, Asn^{102}, Lys^{121}, Asn^{212}, and Tyr^{290} of the human GnRH-R1^{SDP}, known to be important in ligand binding (8).

Phylogenetic Analysis

Phylogenetic analysis of vertebrate GnRH-Rs identified two families of GnRH-Rs (a and b), each containing two subfamilies, which are distinguished by conserved motifs in EC3. The a1 family includes GnRH-Rs from all nonmammalian vertebrate taxa, including fish, reptiles, birds, and amphibians that are characterised by a conserved T*PEY*VH motif in EC3. All mammalian GnRH-Rs, known to regulate reproduction, form a distinct subfamily a2, characterised by a consensus V*SDP*VN motif in EC3 and the absence of a C-terminal tail. With less bootstrap support, family b GnRH-Rs can be split into two subfamilies that are also characterised by different but related motifs in EC3. Teleost and amphibian receptors in subfamily b2 have a V*SHS*LT motif, whereas GnRH-Rs from amphibians and some mammals belong to subfamily b1, characterised by the V*PPS*LS motif. It should be noted that although the GnRH-R2^{PPS} from chicken appears to be an exception, its phylogenetic inclusion in subfamily b2 has a very low bootstrap value. Phylogenetic analysis with alternative algorithms (*e.g.* minimal evolution) yielded very similar topology (data not shown). These and other data (8, 10, 12) showing that GnRH-R families can be distinguished by conserved amino acid motifs in EC3 allow an unambiguous classification system for vertebrate GnRH-Rs. The phylogenetic tree shows that the two types of GnRH-Rs found in teleosts, GnRH-R^{SHS} and GnRH-R^{PEY}, diverged early in vertebrate evolution and are neither recent duplications nor limited to teleosts.

Function of the A. Burtoni GnRH-R1^{SHS}

To discover whether the *A. burtoni* GnRH-R1^{SHS} cDNA encoded a functional receptor, we analyzed its function using a recombinant system. Because fish cell lines were not available, we transfected COS-1 cells with GnRH-R1^{SHS} and showed specific binding of a radiolabelled GnRH analog, which was displaced by mGnRH1 and each of the three forms of GnRH expressed in *A. burtoni*. We found that GnRH-R1^{SHS} had relatively low affinity for the three forms of GnRH that are expressed in *A. burtoni*. IC_{50} values ranged from 53.5 nm for GnRH2, which had highest affinity for GnRH-R1^{SHS}, to 20,300 nm for [Ser^8] GnRH1. Consistent with our previous report, GnRH-R2^{PEY} exhibited similar low affinities for all peptides, and peptides had the same rank order of IC_{50} values in cells expressing either GnRH-R1^{SHS} or GnRH-R2^{PEY} (GnRH2 < GnRH3 < mGnRH1 < [Ser^8] GnRH1).

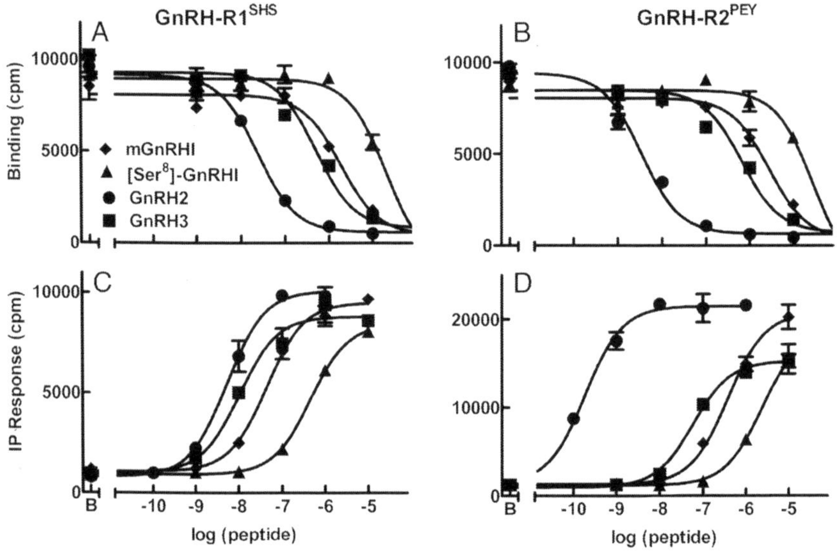

Figure: Competition binding and IP production of GnRH-R1^{SHS} and GnRH-R2^{PEY} receptors. COS-1 cells transiently transfected with cichlid GnRH-R1^{SHS} (A and C) or cichlid GnRH-R2^{PEY} (B and D) were incubated with ^{125}I-labelled [His^5,d-Tyr^6] GnRH (A and B) in the presence of increasing concentrations of mGnRH1 (f&), [Ser^8] GnRH1 (´%), GnRH2 (•), or GnRH3 (ª%) for competition binding assays or (C and D) labelled with [^{3}H]myoinositol and incubated with increasing concentrations of peptides in the presence of LiCl for peptide-stimulated IP production.

The four GnRH peptides stimulated IP production in cells transfected with *A. burtoni* GnRH-R1^{SHS} cDNA, confirming that the cDNA encodes a functional receptor that activates the Gq/11 family of G proteins in COS-1 cells and couples to the IP second-messenger cascade. GnRH2 and GnRH3 exhibited the highest potency (EC_{50} values 8.11 ± 2.63 and 11.0 ± 2.5 nm, respectively), whereas [Ser^8] GnRH1 and mGnRH1 had low potency. In contrast, only GnRH2 exhibited high potency at the GnRH-R2^{PEY} (EC_{50} 0.208 ± 0.034 nm).

In vitro analyses show that the *A. burtoni* GnRH-Rs have similar affinities for each of the GnRH forms in *A. burtoni*. However, comparison of IC_{50} and EC_{50} values indicates that the two receptors differ in their efficiency in coupling binding of different GnRH peptides to intracellular signaling. This is evident from the similar potencies of GnRH2 and GnRH3 (EC_{50} values 8.11 ± 2.63 and 11.0 ± 2.5 nm) at the GnRH-R1^{SHS}, even though GnRH3 has considerably lower affinity

(IC_{50}, 1143 ± 578 nm) at this receptor than GnRH2 (IC_{50}, 53.5 ± 24.6 nm). To quantify these observations, we have calculated an efficiency quotient (IC_{50}/EC_{50}, Table 1Ñ!). Clearly, the GnRH-R1^{SHS} signaling response is particularly sensitive to GnRH3 and poorly responsive to GnRH2, whereas the GnRH-R2^{PEY} is highly responsive to GnRH2. It is notable that neither receptor is highly responsive to [Ser^{8}] GnRH1.

Tissue Distribution and Localisation of A. Burtoni GnRH-R1^{SHS} and GnRH-R2^{PEY}

We mapped the distribution of both GnRH-Rs in *A. burtoni* using RT-PCR for peripheral tissue and *in situ* hybridization for expression in the brain. GnRH-R1^{SHS} is significantly more widespread in the body than is GnRH-R2^{PEY}, being expressed in muscle, intestine, liver, and heart in addition to brain, retina, pituitary, gill, kidney, testes, and ovary where both forms are expressed.

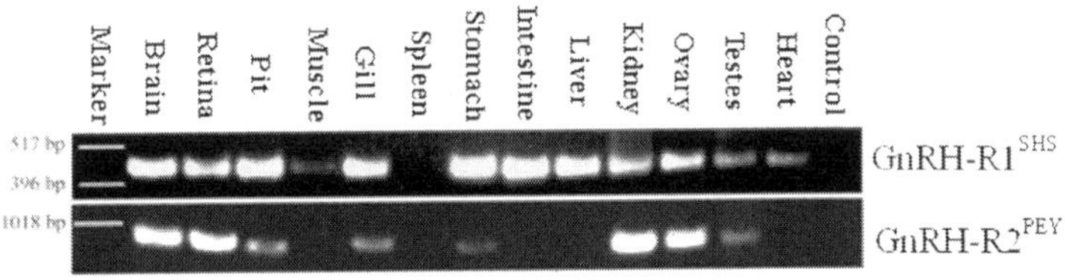

Figure: Tissue distribution of GnRH-Rs. Distribution of mRNA from both GnRH-R types in *A. burtoni* tissue was assessed using PCR. Note that both receptor types are distributed throughout the body, but GnRH-R1^{SHS} is found more widely than GnRH-R2^{PEY}, particularly in the muscle, intestine, liver, and heart. Neither form is found in spleen. The first lane contains markers for 517 and 396 bp for GnRH-R1^{SHS} and 1018 bp for GnRH-R2^{PEY}. Material for the control lane was reverse transcribed without RNA and then used as the PCR template.

To localise GnRH-Rs relative to the gonadotropes and somatotropes, we compared GnRH cell type locations visualized using immunocytochemistry with each receptor type mapped using *in situ* hybridization. The organisation of the teleost pituitary requires that it be sectioned parasagittally to see these relationships clearly. In sagittal sections of the pituitary, the distributions of cells containing GH or LH and a schematic representation are shown. GH- and LH-containing cells are in nonoverlapping regions of the pituitary. Distribution of FSH-containing cells was similar to that of LH-containing cells (not shown). Based on *in situ* hybridization, GnRH-R1^{SHS} is localised in a crescent around the ventral border of the pituitary, and GnRH-R2^{PEY} is located dorsally, directly adjacent to the attachment of the pituitary to the brain, confirming previous results for *A. burtoni* (33). Comparing the *in situ* hybridization results with those of GH and LH distribution in the pituitary of *A. burtoni*, it is clear that the distribution of the GnRH-R1^{SHS} correlates with gonadotropes and GnRH-R2^{PEY} with somatotropes.

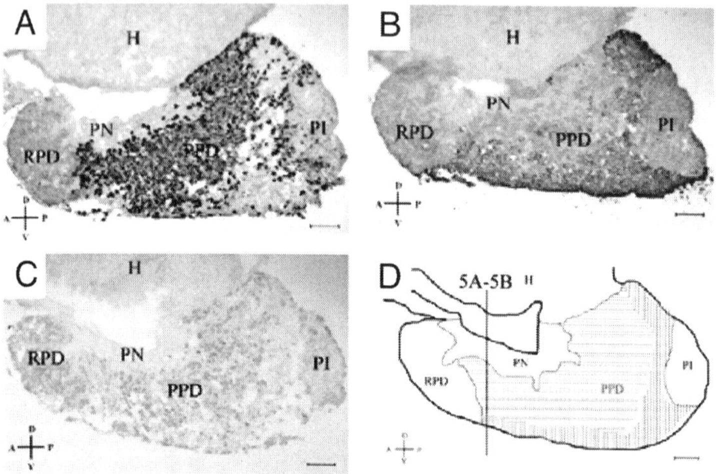

Figure: Localisation of gonadotropes and somatotropes.

Fixed sagittal sections of *A. burtoni* pituitary from T males were incubated with anti-GH antibody (A) anti-LH (B), or no antibody (C). Bound antibodies were visualized with 32 ,32 -diaminobenzidine, and nuclei were counterstained with cresyl violet. A schematic illustration (D) shows the distribution of GH cells (*horizontal shading*) and LH cells (*vertical shading*). The *line* labelled 5A-5B shows the position of the coronal sections of *in situ* staining. *Scale bar*, 100 mm. A, Anterior; D, dorsal; P, posterior; V, ventral; H, hypothalamus; RPD, rostral pars distalis; PN, pars nervosa; PI, pars intermedia.

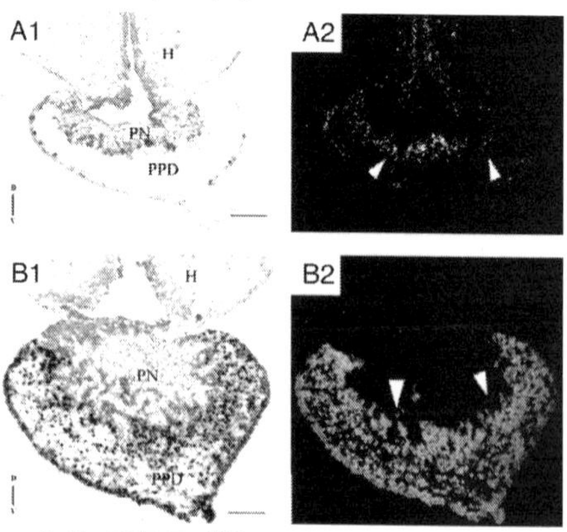

Figure:Localisation of GnRH-R1^{SHS} and GnRH-R2^{PEY} in the pituitary of *A. burtoni* using *in situ* hybridization with ^{35}S-labelled antisense mRNA probes.

Photomicrographs show coronal sections through an *A. burtoni* pituitary showing localisation of GnRH-R2^{PEY} mRNA (A) and GnRH-R1^{SHS} mRNA (B). A1, GnRH-R2^{PEY} mRNA as radioactively labelled silver grains (*black spots*) in bright field with cresyl violet counterstain (*purple*), showing that it is concentrated in the dorsal part of the pituitary in the proximal pars distalis (PPD) adjacent to the pituitary stalk; A2, dark-field photomicrograph of the same field of view where silver grains appear *silver*. *Arrowheads* mark the boundary of the staining that lies between the proximal pars distalis (PPD) and the pars nervosa (PN).

B1, GnRH-R1^{SHS} mRNA as silver grains (*black spots*) in bright field with cresyl violet counterstain (*purple*), showing that it is concentrated in ventral crescent of the PPD; B2, dark-field photomicrograph of the same field of view where silver grains appear *silver*. *Arrowheads* mark the boundary of the staining that lies along the ventral border of the pituitary. *Scale bar*, 200 mm.

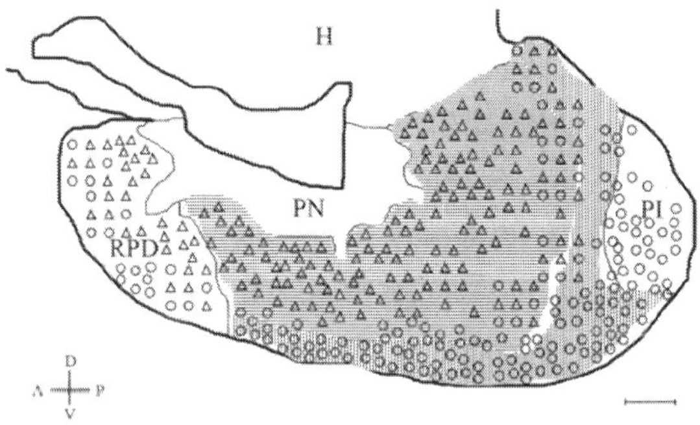

Figure: Schematic illustration of a sagittal section through the *A. burtoni* pituitary showing the distribution of GnRH-R1^{SHS} (Ë%) and GnRH-R2^{PEY} (µ%) receptor types relative to the GH cells (*horizontal lines*) and LH cells (*vertical lines*). *Scale bar*, 100 µm.

Diurnal Rhythm and Coregulation of GnRH-R mRNA Levels

In *A. burtoni*, reproductive behavior in T males and activity in non-T males show distinct daily rhythms (28).

To determine whether there might be a corresponding rhythm at the GnRH-R mRNA level, we collected tissue from three animals at 3-h intervals over a 24-h period and performed quantitative RT-PCR to determine the levels of receptor expression.

Although previous reports suggested GnRH-R mRNA levels change with a diurnal rhythm, we found no statistically significant nonlinearity as would be consistent with a diurnal rhythm in either GnRH-R1^{SHS} or GnRH-R2^{PEY} expression.

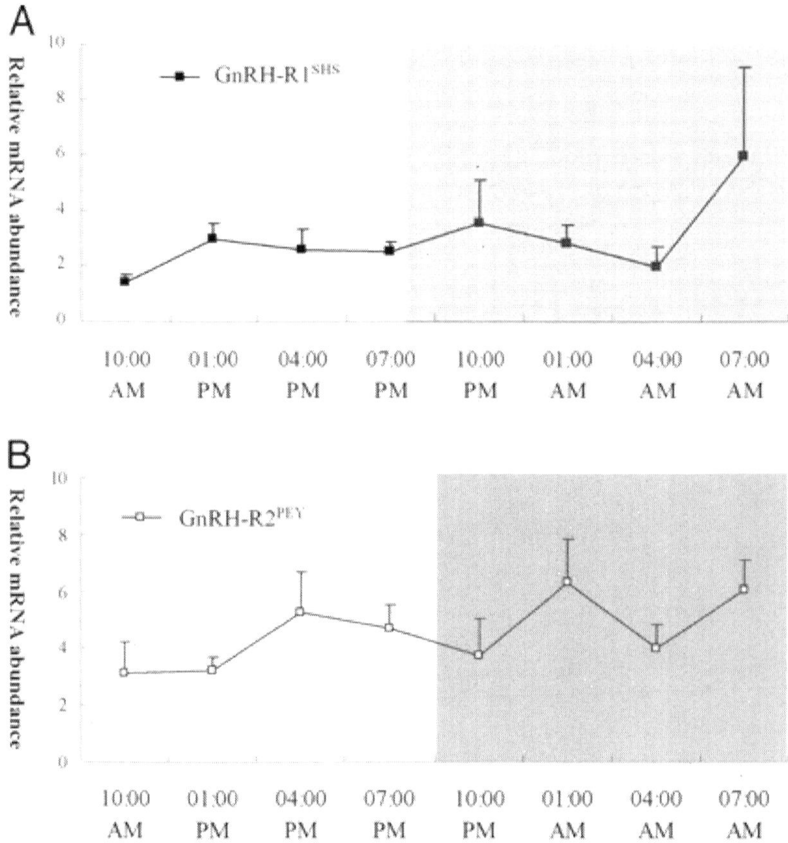

Figure: GnRH-R mRNA abundance as a function of time of day over a 24-h period. Panels show the fluctuation of relative GnRH-R1^{SHS} (A) and GnRH-R2^{PEY} (B) mRNA abundance over 24 h. *A. burtoni* tissue samples were taken at 3-h intervals. mRNA abundance was normalised by mRNA levels of housekeeping genes. *Gray shaded area* represents darkness. *Bars* indicate sem. Lights were on at **0900** h and off at **2100** h (n = 24).

However, we did find significant coregulation of particular forms of GnRH and GnRH-R mRNA in data collected around the clock from the brain and pituitary. There is a highly significant positive linear relationship between each receptor and all three ligands. ICA showed that two independent sources contribute to changes in all the variables in the same direction. This means that when values in the first underlying source's distribution increase, expression of all the mRNAs increases. A third underlying source, however, shows GnRH-R2^{PEY} changing opposite of all other variables. Interestingly, the third underlying source shows GnRH3 is expressed in the same direction as receptor 2 but in the opposite direction of GnRH-R1^{SHS}.

Comparison of Vertebrate GnRH-R Expression and Responsiveness to different GnRH Forms

The physiological role(s) of GnRH-Rs have been assessed by expression in appropriate cells of the pituitary, measurement of expression after physiological perturbations, and measurement of high affinity for GnRH1. To determine whether data in other species support the identification of GnRH-R1^{SHS} as the reproductive receptor in *A. burtoni*, despite its low affinity, we reviewed prior research. We asked whether receptor amino acid sequence correlates with ligand responsiveness and with evidence of expression in gonadotropes or responses to perturbation of the HPG axis in nonmammalian vertebrates. We found no correlation of expression of either SHS- or PEY-type GnRH-Rs with reproductive function.

Because there are limited data on ligand-binding affinities of most vertebrate GnRH-Rs, we compared EC_{50} values. However, absolute EC_{50} values measured in recombinant signaling assay systems are influenced by variables such as receptor expression levels and coupling efficiencies that vary among systems. Because all nonmammalian GnRH-Rs are highly responsive to GnRH2, we compared signaling responses stimulated by various GnRH forms relative to the response stimulated by GnRH2 to calculate a ligand response index. We included all vertebrate GnRH-Rs that fit the four defined receptor types, SDP, PEY, SHS, and PPS, for which we could find dose-response data for GnRH2 and at least one other GnRH form. The comparison shows that ligand response is similar within each receptor type, and there is a strong correlation of receptor sequence with ligand-stimulated response relative to the response to GnRH2. As expected, the mammalian, SDP-type GnRH-Rs are highly responsive to mGnRH1 relative to GnRH2 (mean EC_{50} ratio 0.15), whereas PPS type receptors, the classic type II GnRH2 receptors, show high responsiveness to GnRH2 (mean ratio 239). Of the GnRH-R types that occur in teleosts, the PEY-type receptors are highly selective for GnRH2 and very poorly responsive to all other GnRH forms (mean EC_{50} ratios 140-2900). It is notable that chicken and bullfrog PEY-type receptors are exceptions, exhibiting good responses to GnRH1. In contrast, teleost SHS-type receptors show smaller differentials in peptide-stimulated responses (mean EC_{50} ratios 3.3–58), meaning they are much less selective for GnRH2 and suggesting that a receptor of this type may be the GnRH1 receptor in teleosts.

Cloning and genomic analysis have identified multiple GnRH-Rs in a wide variety of nonmammalian vertebrates, including chickens and several species of fish and amphibians, yet it is not known which receptor type regulates reproduction. The GnRH-R responsible for reproduction must 1) be expressed in the gonadotrope cells of the pituitary, 2) exhibit enhanced expression in animals that are reproductively competent, 3) be responsive to the endogenous GnRH ligand that regulates the HPG axis, and 4) be phylogenetically close to receptors that regulate reproductive function in related species.

To identify which GnRH-R regulates reproduction in *A. burtoni*, we cloned the GnRH-R1^{SHS} receptor, known to be up-regulated in reproductively active T males, and measured its phylogenetic and functional characteristics. We found that GnRH-R1^{SHS} is less closely related than the GnRH-R2^{PEY} to the SDP-type receptors, which are clearly established as reproductive receptors in mammals. Interestingly, neither cichlid GnRH-R has high-affinity interaction with [Ser^{8}] GnRH1, but GnRH-R^{SHS} expression colocalises with gonadotropes, whereas GnRH-R^{PEY} does not.

Phylogenetic Relationships of Jawed Vertebrate GnRH-Rs

Phylogenetic analysis identified four GnRH-R subfamilies with characteristic motifs in EC3. We have defined GnRH-R types according to their EC3 motif, indicated by a three-letter superscript, to reduce confusion of multiple contradictory nomenclature systems for GnRH-Rs and poor correlation of assigned receptor names with receptor type. This unambiguous classification scheme placing GnRH-Rs into four types, SDP, PEY, PPS, and SHS using the EC3 motif, is based solely on receptor amino acid sequence. It has major advantages in that it includes all but a few closely related SHS- and PPS-type receptors and cannot be confused with GenBank assigned receptor names.

It is likely that the four GnRH-R subfamilies identified (a1, a2, b1, and b2, corresponding to types PEY, SDP, PPS, and SHS) arose through gene duplication, an important process in evolution. Duplications allow new gene functions to evolve, and there is good evidence in yeast that duplicated genes can provide functional compensation against null mutations. In this case, GnRH-R^{SHS} and GnRH-R^{PEY} show similar affinity for [Ser^{8}] GnRH1, but they are clearly not functionally redundant, because their expression patterns are divergent. GnRH-R subfamilies can be understood in terms of the two rounds of genome duplication that occurred early in the vertebrate lineage (known as the 2R hypothesis) (53). Because this hypothesis would predict the presence of four GnRH-Rs in all vertebrates, the phylogenetic tree predicts that GnRH-R genes have been lost independently, on multiple occasions from different branches of the vertebrate lineages. For example, SDP-type (family a2) GnRH-Rs have been identified only in mammals and thus must have been lost independently in fish, amphibians, and reptiles. Similarly, PEY-type (family a1) GnRH-Rs are not found in mammals. Amphibians have GnRH-Rs in both subfamilies of family b (PPS-type and P/SQS variations of the SHS type), whereas all fish GnRH-Rs in family b are SHS type and all bird and mammalian receptors are PPS type. Although GnRH-R^{PPS} receptors have been identified in amphibians, chicken, and nonhuman primates, this subtype is absent from the mouse and rat genome sequence data bases. The human GnRH-RII^{PPS} lacks a start codon, is truncated at the second transmembrane segment, and is nonfunctional (54, 55, 56). Similarly, a sheep GnRH-RII^{PPS} ortholog is also nonfunctional (57). Thus, this receptor has been independently lost in rodents and has become a pseudogene in humans and sheep.

Teleost fish have undergone an additional genome duplication since their divergence from tetrapods (58, 59), so it is not unexpected that there are multiple GnRH-Rs of a single receptor type in the *Ostariophysi* lineage including goldfish (GnRH-RA^{PEY} and GnRH-RB^{PEY}) (38), catfish (GnRH-$R1^{PDY}$ and GnRH-$R2^{PEY}$) (25), and zebrafish (GnRH-$R1^{PEY}$, GnRH-$R3^{PEY}$, GnRH-$R2^{SHS}$, and GnRH-$R4^{SHS}$) (12). Similarly, in the *Neoteleostei* lineage multiple GnRH-Rs are present in pufferfish (GnRH-R1/III-1^{SHS}, GnRH-R1/III-2^{SHA}, and GnRH-R1/III-3^{SHS}) and medaka (GnRH-$R1^{SHS}$ and GnRH-$R3^{SHS}$).

In the case of *A. burtoni*, the different expression patterns of the GnRH-R genes suggest that these receptors are in the process of diverging in function. Only GnRH-$R2^{PEY}$ is colocalised on all three groups of neurons that produce GnRH, apparently specialised for feedback control of GnRH production (33). Moreover, GnRH-$R1^{SHS}$ colocalises with gonadotropes, whereas GnRH-$R2^{PEY}$ colocalises with somatotropes, revealing different roles in the pituitary. When detailed data about receptor localisation and function are available from other species, it will become clear whether these specialisations are unique to teleosts and whether evolution has taken different directions in different species.

Role of GnRH-$R1^{SHS}$ as a Functional Receptor for [Ser^8] GnRH1

We have previously shown, in *A. burtoni*, that [Ser^8] GnRH1 is the only GnRH peptide found in the pituitary and that it is up-regulated in reproductively active T males. The low affinity of both *A. burtoni* receptors for [Ser^8] GnRH1 indicates that high concentrations of [Ser^8] GnRH1 (in the micromolar range) would be required in the pituitary to activate either receptor. Superficially, this might suggest that neither receptor is the target of [Ser^8] GnRH1 but that some other receptor with a higher affinity for [Ser^8]-GnRH1 may exist. However, all of the nonmammalian GnRH-Rs that have been functionally analyzed have the highest affinity for GnRH2, and all fish receptors are poorly responsive to GnRH1 forms (11, 12, 25, 27, 38). Available information from zebrafish and pufferfish genomes does not reveal additional GnRH-Rs, making it unlikely that receptors with high similarity to currently known GnRH receptors and high affinity for GnRH1 exist.

Although [Ser^8] GnRH1 had low potency in stimulating IP production in our recombinant system, the efficiency of [Ser^8] GnRH1-stimulated activation of fish G proteins is not known. High-affinity GnRH-R binding has been reported in fish pituitary tissue (60, 61, 62, 63) using synthetic, high-affinity ligands, before the discovery of the teleost forms of GnRH1. However, consistent with results in recombinant systems, catfish GnRH1 was found to have low potency in stimulating gonadotropin release and had very low affinity in GnRH-R binding assays performed with catfish pituitary membranes (60). This suggests the known receptors may regulate reproduction despite their low affinity for the biologically relevant ligands, raising the possibility that synchronous release of GnRH in the pituitary may lead to transiently elevated levels triggering the receptors.

Consistent with their known functions in mammals, all SDP-type GnRH-Rs are highly selective for mGnRH1, and all PPS-type receptors are highly selective for GnRH2. This suggests that PPS-type GnRH-Rs may have a conserved function in amphibians and in nonhuman primates. As in *A. burtoni*, all SHS-type GnRH-Rs exhibit relatively low ligand selectivity, with similar responses to different GnRH peptides, suggesting that SHS-type receptors may have similar functions in teleost and amphibian species. In contrast, all PEY-type GnRH-Rs from fish species are highly selective for GnRH2, suggesting that GnRH2 may be their major physiological ligand. However, PEY-type receptors from chicken and bullfrog are much less selective for GnRH2 and are quite responsive to GnRH1 ([Gln^8] GnRH1 or mGnRH1, respectively) suggesting that the functions of PEY-type GnRH-Rs may differ between teleosts and other vertebrates. Indeed, there is physiological evidence that the GnRH-$R1^{PEY}$ regulates reproduction in chickens, because its expression is up-regulated in the pituitaries of castrated cockerels. However, these data are not consistent with a recent study in which expression of GnRH-R2^{SHS} was shown to increase with reproductive status in male and female chickens (65).

Which GnRH-R Regulates Reproduction in A. burtoni?

Several lines of evidence suggest that the GnRH-R1^{SHS} receptor regulates reproduction in *A. burtoni*. First, GnRH-R1^{SHS} receptor mRNA transcripts are up-regulated in the pituitaries of T males, whereas GnRH-R2^{PEY} receptor mRNA transcripts are not (34). Second, GnRH-R1^{SHS} mRNA transcripts specifically localise with expression of gonadotropes, whereas GnRH-R2^{PEY} transcripts localise with somatotropes. Our results using *in situ* hybridization contrast with results using receptor immunocytochemistry in the closely related teleost *O. niloticus*. In *O. niloticus*, antibodies that would be predicted to recognise the GnRH-R1^{SHS} (AB111356) reacted with somatotrope cells, whereas antibodies that would be predicted to recognise GnRH-R2^{PEY} (AB111357) reacted with gonadotropes.

Because both species are cichlids, this difference is likely to result from the different techniques used. The antibodies were raised against short peptides corresponding to the EC3 sequences of GnRH-Rs. These peptide sequences can be readily identified in diverse proteins by a BLAST search, so it is possible that the antibodies recognise the peptide sequences of other proteins besides GnRH-Rs (57). Thus, although comparing GnRH-R expression patterns in different species does not yield a consensus on whether GnRH-R^{PEY} or GnRH-R^{SHS} receptors regulate release of gonadotropins in nonmammalian vertebrates, our additional expression data confirm that GnRH-R1^{SHS} is the receptor expressed in the gonadotropes of *A. burtoni*.

Coregulation of GnRH Ligand and Receptor Expression

The widespread distribution of GnRH-Rs (66, 67) has led to speculation that different forms of GnRH may play roles beyond reproduction. Chen and Fernald recently showed that the receptor types differ in their brain distribution in *A. burtoni* and only GnRH-R2^{PEY} colocalised with the three GnRH-producing cell types in the brain, suggesting direct feedback control of the GnRH production. The coregulation of receptors with GnRH suggests that the GnRH system has some common regulatory underpinnings. The presence of multiple GnRH forms across all vertebrates suggests they might share some functional role in reproduction with GnRH1.

GnRH-R Evolution and the Usefulness of the EC3 Motif Classification System

The poor correlation of GnRH-R sequence with a role in reproduction across species makes it impossible to predict which GnRH-R will regulate reproductive function in any particular species. The absence of SHS-type GnRH-Rs in higher vertebrates precludes their role as the receptor responsible for reproduction in these species. Different families of duplicated receptor genes regulate reproductive function in mammals (family a, SDP-type) compared with fish (family b, SHS-type). There is insufficient evidence in other taxa (birds, reptiles, and amphibians) to conclude which receptor is used or when selection for one receptor rather than another might have occurred. Thus, it makes sense to define GnRH-R types by genotype, which, perforce, correlates with EC3 amino acid motif and with ligand selectivity, rather than using a classification system based on physiological function in the face of insufficient and conflicting information.

In summary, we cloned the GnRH-R1^{SHS} receptor, which is up-regulated in reproductively active *A. burtoni*, and showed that it is coexpressed with LH in the pituitary. Like all teleost GnRH-Rs, it binds and responds poorly to GnRH1. Although this remains puzzling, it suggests that synchronous release of GnRH1 into the pituitary allows delivery of sufficient ligand to activate the receptor. The absence of the SHS-type GnRH-R in many taxa and the poor correlation of GnRH-R sequence with a role in reproduction across species, suggest that different families of duplicated GnRH-R genes have been selected to regulate reproduction during vertebrate evolution.

Reproductive Physiology of Fish: a 30 Year Overview

Research into the physiology of fish has had many beneficial implications for the aquaculture industry, explains Bernard Jalabert, Fish Reproduction Group. This feature was published by Cybium, Société Française d'Ichtyologie.

Introduction

The first International Symposium on the Reproductive Physiology of Fish was held 30 years ago, in September 1977, at the Biological Research Station of Paimpont, in Brittany. It was the result of a pioneer initiative by Roland Billard, head of a research group on fish reproductive physiology, which was founded a few years ago in Jouy-en-Josas (France). The symposium gathered approximately 80 participants, and oral presentations were published in a special issue of the *Annales de biologie animale, biochimie, biophysique,* Vol. 18(4). Most of them were basic studies concerning mainly gonadotropins and gametogenesis, with a few devoted to direct applications for aquaculture (Tab. I).

Reproductive physiology of finfish (RPF) is a vast scientific field, which directly concerns a set of physiological processes essential for reproduction, beginning with egg fertilization and ending with sexual behaviour and spawning. Processes include gonad differentiation, puberty, male and female gametogenesis, and the timing of reproductive cycles, all regulated by numerous neuroendocrine, endocrine, paracrine and autocrine factors, all along the brainpituitary- gonadal (BPG) axis. These processes also interact with other important physiological functions such as growth (Le Gac *et al.,* 1993), nutrition (Izquierdo *et al.*, 2001), osmoregulation (Haffray *et al.*, 1995; Le Francois and Blier, 2003), and response to stress factors (Schreck *et al.*, 2001)... RPF is also highly dependent on external factors such as temperature, photoperiod and water salinity, which may modify the various stages of the reproductive process, depending on the species. In fact, fish exhibit a huge variety of specific reproductive strategies and tactics (Wooton, 1984), with just as many specific adaptations in terms of physiological regulation.

It is now well established that scientific knowledge in the RPF field has numerous applications. This is obviously the case for aquaculture production of many species reared for human consumption, with applications such as hormonal spawning induction (Billard, 1993; Zohar and Mylonas, 2001), off-season spawning (Bromage *et al.*, 2001), broodstock and gamete management (Billard, 1992; Bromage *et al.*, 1992), sex ratio control (Baroiller *et al.*, 1999), and age at puberty (Okuzawa, 2002). The same kind of innovations might also be useful for the production of aquarium pet species, thus reducing collection pressure in the wild. In both cases, diversification of species is generally sought thus justifying the development of comparative research on reproductive strategies and physiological regulations. Preservation of domesticated and wild genetic resources, through gamete storage and/ or embryo regeneration, is another research area (gamete physiology) with current and potential applications... Fisheries management might also take advantage of knowledge concerning the reproductive potential of species in order to improve modelling of exploited populations. Finally, the recognition of the sensitivity of fish reproductive processes to the presence of xenobiotics in the aquatic environment (Jalabert *et al.*, 2000) has been widely utilised in the development of a variety of endocrine disruption biomarkers (Hutchinson *et al.*, 2006). From a more general point of view, it appears that finfish are appropriate models not only from the perspective of ecotoxicological risks, but also in terms of comparative biology, due to the significant similarity of basic aspects of the BPG axis among vertebrates (Blazquez *et al.*, 1998).

Development of research in the RPF field during the last three decades cannot be dissociated from the concomitant evolution of the general socio-economic context. According to FAO data, the world finfish aquaculture production has grown from around 3.6% of total finfish production in 1975 (58.4 million tons), to 27.6% of total production (about 110 million tons) in 2005. In addition, since capture fisheries have been stagnating for the last 15 years, the world fisheries production increase during this period is due entirely to aquaculture production development. This increase is driven by both the increasing world demand for fish (which capture fisheries can no longer meet due to overexploitation of natural resources), and by the application of new biotechnology in all fields of fish biology, in particular reproductive biology. An increase in public concern has also been observed during this period, driven by fears about the possible negative health and environmental impacts of new production technologies, and expressed through a growing demand for sustainable production methods.

Since the publication of an exhaustive review of knowledge in the field of RPF 26 years ago (Sundararaj, 1981), the number of scientific papers published each year in the field has increased tremendously. Therefore, the present article does not seek to be exhaustive, but will focus on a few significant discoveries in selected areas of RPF: gonadotropins and gonadotropin releasing hormones (GnRH) as key regulators of reproductive function; and puberty as an example of a reproductive function process dependant on specific reproductive strategies, exhibiting various reproductive tactics and regulated by a cascade of endocrine, paracrine and autocrine factors. Finally, some examples of economically relevant applications of RPF knowledge will be recalled.

Fish Gonadotropins: from one GtH in 1977... to FSH and LH Today

In 1977, relatively purified gonadotropin (GtH) preparations had been extracted from carp, salmon, sturgeon and trout pituitaries through biochemical separation techniques. Some of them were already sufficiently homogeneous for the purposes of radio-immuno assays in endocrinological studies. However, even after validation of purification steps by global biossays typical of final gametogenesis steps, such as spermiation or intrafollicular oocyte maturation (Breton *et al.*, 1976), there was still no clear biochemical or biological evidence of gonadotropin duality in fish. Besides, cytological observations of pituitary cells had suggested the presence of one or two cell types in teleost fish, depending on the species. Yet, the first isolation of two types of gonadotropins, exhibiting predominant vitellogenic or maturational activities, was achieved in the plaice (Ng and Idler, 1978). Later, two distinct salmonid GtH, GtH I and GtH II, were isolated and characterised by their respective steroidogenic potency in vitro (Suzuki *et al.*, 1988; Swanson *et al.*, 1991). Since then, both GtHs have been prepared in many other species, through classical purification methods and/or, more frequently, sequencing and homology cDNA cloning: common carp, grass carp, eel, killifish, striped sea-bass, channel catfish, European sea-bass, red sea-bream, tilapia, Atlantic halibut, grouper (Li *et al.*, 2005; Yaron *et al.*, 2003). Although the structural homology of the â subunit of a purified GtH from chinook salmon to human LH was already suggested twenty years ago by partial protein sequencing and cDNA cloning (Trinh *et al.*, 1986), the identification of GtH I and GtH II to FSH and LH respectively was finally formally adopted at the 6th ISRPF in Bergen in 1999. The characterisation and availability of both gonadotropins have allowed specific assays to be set up and experiments at endocrinological, cellular and molecular levels to be carried out, thus leading to significant progress in the understanding of their respective physiological targets and mechanisms of action.

Regulation of Pituitary Gonadadotropin Secretion and Multiplicity of GnRH Functions

Most data in the field were obtained either at the time of oocyte maturation and/or by using radio-immuno assays of GtH I (LH), the first gonadotropin available in goldfish and salmonids. Therefore, the terms "gonadotropin" or GtH in the following paragraph will refer implicitly to LH.

In 1977, the presence in the fish hypothalamus of a LHRH- like factor able to promote gonadotropin release had already been shown a few years ago (Breton *et al.*, 1971). Mammalian synthetic LH-RH could therefore be used to observe variations of GtH secretion response in vivo, dependent on the gametogenetic stage (Weil *et al.*, 1978). In fact, the 1st fish GnRH, sGnRH was characterised only a few years later, after purification from chum salmon brain (Sherwood *et al.*,

1983). In the meantime, the inhibitory effects of dopamine on gonadotropin release, induced by injections of a luteinizing hormone-releasing hormone analogue (LHRHa), was demonstrated in female goldfish (Chang and Peter, 1983). From that time on, the model of gonadotropin secretion regulation was progressively enriched by numerous data. First, it had to take into account that GnRH was not only originating from the hypothalamus but also from other brain areas (Kah *et al.*, 1993). Moreover various brain neurotransmitters, together with steroid feed-backs, appeared able to modulate the main stimulating/inhibitory roles of GnRH/dopamine, thus relaying the action of external factors. A general model of regulation was thus published (Peter and Yu, 1997), in which GnRH is identified as the main stimulatory factor in all species investigated, whereas the inhibitory role of dopamine is more or less important, depending on the species.

In addition to the control of gonadotropin release by the pituitary, GnRH has also been involved in the control of GH release in the goldfish pituitary (Marchant *et al.*, 1989). More recently, due to the accumulation of new data favoured by the availability of molecular tools, it emerged that GnRH exhibits a large functional diversity, not only in fish, thus again boosting new comparative research in all vertebrates (Lethimonier *et al.*, 2004; Sherwood and Adams, 2005). Data in fish concern: the multiplicity of natural variants (at least 8 variants among fish species investigated so far, with 2 to 3 variants in each species); the multiplicity of receptors (GnRHRs), each one exhibiting different affinities to different GnRH natural variants or analogues; the multiplicity of sites of expression (not only in the brain, but also in the gonads, gills, liver, kidney and other tissues, depending on the species); and the multiplicity of biological effects, such as cell proliferation and steroidogenesis in males and meiosis reinitiation in females.

At the pituitary level, the development of a novel single cell real-time quantitative PCR technique has shown that multiple combinations of three GnRH receptor types could be differentially expressed in individual gonadotropic and non gonadotropic cells of tilapia *Oreochromis niloticus*, thus demonstrating cellular and functional heterogeneity in correlation with reproductive state (Parhar, 2005). All the above data, together with comparative data in other animals, suggest that pleiotropic GnRH function in fish might result from evolution of an ancestral gonadotropic function in primitive organisms (Kah, 2004).

Puberty in fish: an example of a reproductive trait important for aquaculture production, exhibiting a variety of specific strategies and tactics, modulated by external factors, and under complex physiological control. Puberty designates both the transition period and the underlying physiological process corresponding to the acquisition of the capacity to reproduce. In fish, puberty is generally considered as beginning with the apparition of the first spermatocytes in the male, and the initiation of vitellogenesis in the female (Le Bail, 1986), although the second criteria is controversial. Puberty is one of the main traits representative of specific reproductive strategies and tactics as defined by Wootton (1984): each species exhibits a general strategy regarding the age at puberty, such as late puberty in sturgeons or early puberty in tilapias. Depending on various environmental factors, most species are also able to exhibit adaptive tactics within this general strategy. For example, this is demonstrated by the case of precocious puberty in the sturgeon under artificial farming conditions (Doroshov *et al.*, 1997), by precocious male puberty in salmonids, and early puberty in tilapia induced by a protein-enriched diet *Oreochromis niloticus* (Gunasekera *et al.*, 1995).

Early puberty, which is observed in most farmed species, diverts much metabolic effort towards development of the gonads, although it is often preceded or accompanied by accelerated body growth that may be a desirable trait. This paradox is an important problem in fish farming, partly resolved through food restriction or control of external factors such as temperature or photoperiod. It brings to the fore the existence of a subtle physiological equilibrium between growth and puberty, modulated by external factors.

Concerning the possible internal factors determining puberty, the "gonadostat" and "missing link" concepts (Schulz and Goos, 1999) present some heuristic interest. The "Gonadostat" refers to the negative steroid feedback upon the GnRH system in mammals before puberty. The "missing link" assumes that one or more components of the Brain-Pituitary-Gonad axis are non-functional before puberty. Information has progressively been acquired through several approaches. The stimulatory effect of exogenous steroids on the gonadotropic activity in immature male fish has been demonstrated in various species, both at the brain and/or pituitary and/or blood and gonadal levels (rainbow trout (Atteke *et al.*, 2003; Magri *et al.*, 1985), platyfish (Schreibman *et al.*, 1989), African catfish (Cavaco *et al.*, 1998), seabass (Zanuy *et al.*, 1999), (Okuzawa, 2002). This rules out the "gonadostat" concept, according to Schulz and Goos. However, whereas this effect can be accompanied by the onset of spermatogenesis in males, no stimulation of vitellogenesis has ever been reported, suggesting that a gonadal steroid is not the only "missing link", at least in females.

Concerning the role of pituitary hormones, many feed back experiments based on the administration of exogenous steroids have given contradictory results regarding the respective role of FSH and LH, especially in salmonids (reviewed in Okusawa, 2002). However, molecular studies at the pituitary level and endocrinological studies carried out throughout puberty suggested that FSH, and to a lesser extent GH, could be involved (preferentially to LH) in spermatogenesis and vitellogenesis initiation in prepubertal rainbow trout (Gomez *et al.*, 1999). In addition, a consistent positive relationship between plasma IGF1, estradiol-17beta, and pituitary FSH was observed during the growth of coho salmon in spring, suggesting that these factors provide an important link between body growth and oogenesis (Campbell *et al.*, 2006).

The possible involvement of leptin, a peptidic hormone secreted by adipocytes, in the control of puberty through regulation of gonadotropins or somatolactin secretion was recently investigated respectively in trout (Weil *et al.*, 2003) and sea bass (Peyon *et al.*, 2003). In trout, the absence of the response of pituitary cells from immature fish to recombinant human leptin, and the stimulation of FSH and LH release only when gametogenesis was fully engaged, suggested that leptin could not be the unique trigger for puberty. Besides, the positive response of somatolactin secreting cells from prepubertal sea bass to mouse recombinant leptin suggested that leptin might play a role in this species, through growth regulation by somatolactin.

At the brain level, the recent discovery in mammals of the key role of kisspeptin and its protein-G coupled receptor (GPR54) in the control of puberty (Messager *et al.*, 2005) was rapidly followed by GPR54 cloning and detection in individual GnRH neurons and pituitary cells of tilapia (Parhar, 2005). Interestingly, the latest techniques allowing mRNA expression profiling at the individual cell scale also demonstrated that multiple GnRH receptors (GnRH-Rs) are co-expressed, not only in lactotropes, somatotropes, luteinizing and follicle stimulating hormone cells, but also in somatolactin, thyrotrope, corticotrope and melanotrope cells (Parhar *et al.*, 2005). Since then, expression patterns of GPR54 mRNA in the brain of grey mullet (Nocillado *et al.*, 2007) and cobia (Mohamed *et al.*, 2007) were observed as being concomitant with puberty onset.

In conclusion, new endocrinological and molecular evidence confirms the close inter-relations between puberty and growth regulation at all levels of the BPG axis. However, the ultimate trigger mechanism for the initiation of puberty remains unclear. Kisspeptin certainly appears to be a key part of that mechanism, but what controls its release and how? Recent research often uses a "candidate gene approach". However, the present availability of large-scale, and high-throughput methods, in the functional genomics of whole tissue transcriptome provide hope for new discoveries. For example, it was recently observed that 15% of the 3000 genes surveyed in brain transcription profiles of early maturing Atlantic salmon "sneakers" males, immature males and immature females exhibited differential expression (Aubin-Horth *et al.*, 2005). Pertinent candidate genes for puberty onset might be expected among these.

Concluding Remarks Concerning the Application of Fish Reproductive Physiology Research to Aquaculture

Research in RPF represents a large scientific community all over the world, sustained by public, economic and environmental concerns in many countries. Scientific approaches concern various complementary levels, from basic knowledge to directly applied research.

In many cases, practical demand, originating mainly from the fish farming sector, has been pushing research to find solutions for the control of reproductive traits of economic interest, almost independently of research into underlying physiological regulations. This has been the case concerning manipulation of external factors, such as photoperiod or temperature. Thus, photoperiod manipulations have been applied in many farmed species to control out-of-season spawning (Bromage *et al.*, 2001) and puberty (Okuzawa, 2002). In both cases, however, precise protocols depend not only on the species but also on the genetic strain: in rainbow trout for example, the same photoperiod regime cannot be applied to an autumn-spawning and a spring-spawning strain. Moreover, interactions with other factors are either unknown or cannot be easily taken into account for practical reasons. This is the case for temperature, for which specific requirements are not limited to the spawning period, thus impairing egg quality, as shown in Arctic char (Gillet, 1991), rainbow trout (Davies and Bromage, 2002), Atlantic salmon (King *et al.*, 2007), or halibut (Brown *et al.*, 2006). Besides, inadequate photoperiod manipulation by itself can also impair egg quality, as in trout (Bonnet *et al.*, 2007) or in Eurasian perch (Migaud *et al.*, 006). In fact, the manipulation of external factors to control any reproductive step is generally limited by the ignorance of the precise mechanisms by which these factors act, and interact, on underlying physiological processes. As already underlined in the case of puberty, much can be expected from largescale and high-throughput methods applied to functional genomics of whole tissue transcriptome and proteome, to understand multiple interactions between external factors and physiological regulations.

Applications of RPF knowledge generally bring progressive improvements to aquaculture. This is the case, for example, of hormonal spawning induction. Initially a solution was provided in the form of injections of crude pituitary extracts, which were successively replaced by partially purified gonadotropin, then by various GnRH analogs, sometimes in combination with anti-dopaminergic compounds (reviewed by Zohar and Mylonas, 2001). However, rapid industrial development of aquaculture may follow on from the development of techniques that unlock a reproductive block, as in *Pangasius* fish farming in Vietnam. The annual farmed production of this species, based on fry capture in the wild, was stagnating at around 12500 tons until 1995. Then, the demonstration that successful maturation and ovulation could be artificially induced by serial hCG injections (Cacot *et al.*, 2002; Cacot and Lazard, 2004), boosted annual production up to 440000 tons in 2005 and probably near 1 million tons in 2007. Most of the increase concerns *Pangasius hypophthalmus* species, initially considered as less valuable with regards to flesh quality. This apparent paradox results from higher fertility and easier induction treatment of *Pangasius hypophthalmus*, but it also points to the essential limiting factors of species domestication, such as fertility (number of eggs per fish) and practicability of possible treatments in field conditions.

Finfish aquaculture based on seed capture in the wild, or even on the capture of adults, still concerns species such as eels, tunas, groupers and yellowtails, some of which are now considered as endangered species. Therefore, research on RPF has still much scope for progress, not only to remove the pressure on the species mentioned above, but also to find new candidate species for aquaculture by increasing efforts in the field of comparative reproductive biology of new species.

Finally, it may happen that advances in RPF knowledge open the door to innovative techniques that do not result of any demand from farmed fish producers and that can even be rejected by consumers in some countries, like transgenic fish for specific characteristics such as improved growth or disease resistance. In this field particularly, but also in the case of any innovation, public concern about biotechnological applications and sustainable aquaculture development must be taken into account.

Studies on the Reproductive and Developmental Biology of Cichlasoma Dimerus *(Percifomes, Cichlidae)*

ABSTRACT: Many characteristics of the South American teleost fish *Cichlasoma dimerus* (body size, easy breeding, undemanding maintenance) make it amenable to laboratory studies. In the last years, many of the fundamental aspects of its reproductive and developmental biology have been addressed in our laboratory. Rather recently, the immunohistochemical localisation of pituitary hormones involved in reproduction and in background colour adaptation has been described in both adult and developing individuals, and the role of FSH in ovarian differentiation has been established. These findings have been correlated with mapping of some of their brain-derived controlling hormones. The latter include brain-derived gonadotropins which were shown to be active *in vitro* in the control of pituitary hormone secretions. The emerging picture shows *C. dimerus* as an interesting species in which many of their basic features have already been investigated and which conform a solid platform for comparative studies correlating neurohormones, pituitary hormones and behavior, from the molecular to the organismic level.

Teleost Fishes in Physiological and Developmental Studies.

Teleost fishes represent the most numerous and diversified vertebrate group. Though fish physiology has been studied for many decades (Ball and Baker, 1969; Hoar, 1969; Conlon, 2000; Price *et al.*, 2008), the use of very different species, together with the idea that differences were more important than similarities, gathered a lot of data, and sometimes non coherent ones. During the last years, however, several research groups have put the emphasis in the importance of using teleost fishes not for comparative studies within the class, but as general vertebrate physiological models, by taking advantage of some of their unique properties. One of these properties is an unusual characteristic of the fish nervous system: fish brains can remain alive and responsive for many hours after removal from the animal. This has allowed researchers to use intactbrain preparations that have the advantage, compared to mammalian brain-slice preparations, of allowing afferent and efferent connections to remain functional (Wayne *et al.*, 2005). Another interesting feature is the morphological arrangement of fish neuroendocrine systems: in teleost fishes there is no median eminence or portal system; therefore nerve fibres originating from the hypothalamus should end in close contact with pituitary cells. Immunocytochemical studies have shown that neurohormonal fibres and nerve endings are in close association with pituitary endocrine cells. Because of this special anatomical arrangement, teleosts are a unique experimental model for determining putative brain peptides or monoamines involved in the regulation of pituitary endocrine cells (Peter *et al.*, 1990).

Also, over the past 15 years, the cyprinid zebrafish *Danio rerio* (Hamilton, 1822) and the adrianichthyid medaka *Oryzias latipes* (Temminck and Schlegel, 1846) have become very popular fish models, due to their transparency during embryonic and larval stages, for investigating the molecular genetics that control organ development. The by-product of this huge amount of work is that their genes have been characterised providing fundamental tools for physiological and developmental studies (Furutami-Seiki and Wittbrodt, 2004; Crollius and Weissenbach, 2009).

Cichlid fishes (Perciformes) are also becoming intensively studied in the field of social control of reproduction and the most studied species have been the African tilapia *Astatotilapia burtoni* (Günther, 1894) and *Oreochromis mossambicus* (Peters, 1858) (Fox *et al.*, 1997; Ogawa *et al.*, 2006). The South American cichlid fish *Cichlasoma dimerus* (Heckel, 1840) has been studied for several years in our laboratory, at the morphological, developmental, physiological, and molecular levels. The aim of the current review has been to present a comprehensive picture of all the data obtained in this species, and to put them in the context of contemporary, comparative fish physiology.

Cichlasoma dimerus in the laboratory and in nature.

It is a medium sized fish (12 cm standard length) which is easily maintained and bred in the laboratory, where it tolerates a wide range of water compositions and temperatures (10 to 30°C, the optimal breeding temperature is 26°C) (Meijide and Guerrero, 2000).

The natural range of this species encompasses the entire system of the Paraguay river, the lower Alto Paraná, and the rest of the Paraná river basin up to the vicinity of Buenos Aires city (Kullander, 1983). Locality records are known from four countries (Bolivia, Brazil, Paraguay and Argentina) where it inhabits a wide variety of both lentic and lotic environments. Its common names are "chanchita" (Spanish) and "acará" (Portuguese) (Staeck and Linke, 1995; Casciotta *et al.*, 2005).

Figure: Some behavioural aspects of Cichlasoma dimerus under laboratory conditions. A: Fin display and body colours of an adult male of high social rank. B: Prespawning behavior: the female nibbles at the spawning substrate (a flat stone), while the male is on guard. C: Spawning behavior: the female rubs the genital papilla while depositing the last egg rows. D: Parental care of the clutch: the female is removing unfertilized or infected eggs. Scale bars = 20 mm.

Its ground colour is very variable, depending both on the specimen's "mood" and social status. Its body may vary from greenish to light or dark grey and it may also show golden-yellow and light blue reflections. It shows several dark-brown vertical bands and two blotches, one in the middle region of the trunk and the other in the caudal peduncle. The eyes may show a bright red border, particularly in reproductively active individuals. This species has a moderately developed sexual dimorphism, with the males growing larger than the females. In males, soft rays of the distal edge of the dorsal and anal fins may be extended as filaments (Alonso *et al.*, 2007).

As many cichlids do, *C. dimerus* has highly organised breeding activities, as they may be observed in the laboratory. A few days after fish have been transferred to a new aquarium, territories are progressively established and defended by both males and females. The dominant pair will aggressively defend the prospective spawning site (usually a flat stone) and will start to display stereotyped prespawning activities, such as jerking and quivering, digging in the gravel, and nipping off and nibbling at the spawning substrate.

Spawning will eventually initiate, with the female rubbing her genital papilla over the substrate and the sticky eggs being deposited in rows. After the female lays its eggs, the male slides its erected genital papilla over them, thus achieving fertilization. At the end of the spawning process, which may last up to 90 min, some 1500 eggs have been deposited as a uniform rounded layer over the substrate.

Spawning is followed by a period of parental care during which the cooperative pair guards the eggs, both by fanning them (alternately beating the pectoral fins over the clutch) and by removing dead eggs, which are usually attacked by fungi. At a water temperature of 26°C, the larvae hatch at the beginning of the third day and are transferred by both parents to a previously dug pit. Another five days pass until the larvae swim freely, and during this time they are recurrently transferred to several other pits. Mortality during the first month of development is typically high (only about 20% of hatchlings reach the juvenile stage). If provided with enough food and space, the young will grow rapidly, and may reach about 50 mm four months after hatching. A pair may spawn every 20 days during the extended breeding season (September to April).

The foregoing behavioral description was based on the works of Greenberg *et al.* (1965) and Polder (1971) in *Aequidens portalegrensis*, which is a synonym of *C. dimerus*. A comprehensive monograph on cichlid behavior by Baerends and Baerends-Van Roon (1950) is also available.

Embryonic and Larval Development.

The freshly spawned eggs of *Cichlasoma dimerus* are ovoid-shaped and light brown coloured (Meijide and Guerrero, 2000). They have a smooth translucent vitelline envelope surrounded by a mucous layer which allows them stick to the substrate and to one another at the time of spawning.

A detailed description of *C. dimerus* development from oocyte activation until the beginning of the juvenile stage is found in Meijide and Guerrero (2000). The micropyle is cone-shaped and located at the egg's animal pole. Most of the egg volume is occupied by numerous bright, tightly packed yolk globules. Polarization of the cytoplasm towards the animal pole becomes evident within 60 min after fertilization, giving rise to a prominent dome-shaped cytoplasmic layer known as the blastodisc.

Cleavage is typically meroblastic, resulting in a blastula with a blastodermal cellular region above the yolk mass. The subsequent gastrulation process is characterised by several morphogenetic movements that result in a rearrangement of the blastoderm relative to the yolk. Towards the end of the first day of development, about 75% of the yolk mass has been covered by the blastoderm and the elongated embryo's shape is discerned along an anterior-posterior axis. The "yolk plug" is the portion of uncovered yolk protruding in the neighbourhood of the vegetal pole.

Organogenesis initiates soon thereafter, the optic vesicles being one of the first organ primordia that become distinct. Towards 26 h, paired somites start developing sequentially in an antero-posterior direction, on both sides of the notochord. Otoliths become visible within the otic vesicles. Before hatching, the embryo shows conspicuous muscular contractions. The beating heart is placed beneath the head and blood flow is readily observed. Brain regions become distinct and the lenses are evident within the eyes. Rudiments of the adhesive glands are observed at the midbrain-hindbrain boundary.

Larval hatching occurs 54 h after fertilization. After hatching, yolk-sac larvae are readily transferred by their parents to a pit dug in the gravel. These larvae present three pairs of adhesive glands over the head, this feature being characteristic of substrate-brooding cichlids. The mouth is not yet formed, and the gut is narrow and straight. No fins are differentiated but a primordial fin fold is already developed dorso-ventrally in the sagittal plane. Melanophores are scattered over the oblong yolk-sac and extend along the junction of the body with the ventral fin fold.

During the fifth day of development, the mouth opens and mobile jaws can be observed. The yolk-sac has become smaller. Branchial arches are differentiated and soon thereafter gill filaments will start to develop. These larvae do not swim freely yet, but they rapidly swing their tails while still attached to the substrate by their adhesive glands.

The free-swimming stage is reached five days after hatching, i.e. during the eighth day of development. The oval-shaped, refractive swim bladder becomes visible. Adhesive glands have been regressing, and will completely disappear during the following days. Caudal fin rays are well developed, while while both dorsal and anal fins consist of thin, transparent membranes. Melanophores are more heavily distributed over the head and trunk. A few hours after the onset of free-swimming, these larvae start to feed from the environment, though the yolk is not yet totally consumed.

Larval development continues with the progressive differentiation of the dorsal, anal and pelvic fins and with the acquisition of a pigmentation pattern that gradually resembles that of the adult. The juvenile stage is reached around day 45 of development and their body shape and pigmentation pattern are similar to those of the adult. The fins become fully developed and the adult complement of rays is completed. The scales are present over the greatest part of the body, which is almost completely covered with melanophores.

Gonadal Development and Sex Differentiation

C. dimerus is a gonochoristic fish, in which ovaries and testes develop directly from undifferentiated gonads. A detailed description of early gonadogenesis in *C. dimerus*, including ultrastructural features, is found in Meijide *et al.* (2005). As it is known for other teleosts (Bruslé and Bruslé, 1983), the extent of gonadal development in *C. dimerus* depends both on age and size, and gonadal sex differentiation will occur earlier in specimens that grow faster. Therefore, the ages corresponding to the developmental events described in the next paragraphs should be considered only as estimations.

On day 12 posthatching, paired gonadal primordia are present in the posterior region of the abdominal cavity, immediately below the kidney, and are suspended from the dorsal peritoneal wall by short mesenteries. The gonadal primordium consists of large, round to oval germ cells surrounded by somatic cells. Germ cells contain large and central euchromatic nuclei, with prominent nucleoli. Occasional mitotic activity is observed in these cells. Somatic cells are of varying shapes. They have heterochromatic nuclei and a small cytoplasm. Gonads remain undifferentiated by day 38. However, the number of germ cells has augmented, and as a result, gonads increase in size and several germ cells (not just one or two as in earlier stages) appear in cross sections.

Ovarian differentiation occurs at an earlier age than that of testicular differentiation, as it is usually observed in teleosts (Meijide *et al.*, 2005). A prominent cytological feature marking the onset of gonadal differentiation is the appearance of meiotic figures in germ cells of the prospective ovary. By day 42, many oogonia have entered meiotic prophase and become growing primary oocytes, which in addition to a moderate oogonial and somatic cell proliferation result in an enlargement of the prospective ovary. For the first time, blood capillaries are observed in the gonadal dorsal region. The onset of ovarian meiosis is soon accompanied by an increase in germ cell mitoses. In addition, the ovary becomes either triangular or kidney-shaped in sections. A few days later, somatic reorganisation of the developing ovary is evidenced by somatic cell proliferation, which forms appendix-like structures which finally fuse to form the ovarian lumen.

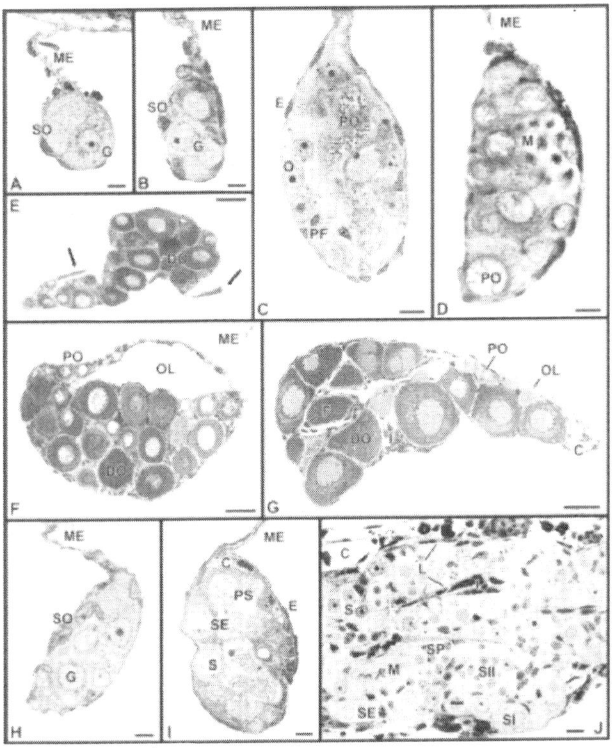

Figure: Gonadal development and sex differentiation in Cichlasoma dimerus. Toluidine blue. A: Primordial gonads on day 14 post-fertilization are characterised by the occurrence of a few germ cells with round nuclei, which are surrounded by some somatic cells. B: Sexually undifferentiated gonad on day 38. C: The appearance of meiotic figures is the first evidence of sexual differentiation in prospective females on day 42. D: Increased gonial mitotic activity accompanies the first cells undergoing meiosis towards day 50. E: Arrows indicate somatic cell outgrowths in a female on day 50, which will lead to the formation of the ovarian cavity. F: The ovarian cavity is already formed on day 65. Numerous basophilic diplotene oocytes are also observed. G: Ovarian follicles containing perinucleolar oocytes are observed by day 80. H: An undifferentiated gonad on a day 65 juvenile (a prospective male). I: Meiotic figures indicative of testicular differentiation are first seen on day 80 in a prospective male. J: Lobules are becoming distinct in a testis on day 100. Secondary spermatocytes (but no spermatids) are observed. Scale bars = 5 µm (A, B, H, I); 10 µm (C, D, J); 50 µm (E, F, G).; C, blood capillary; DO, diplotene oocyte; E, epithelial cell; F, follicle cell; G, germ cell; I, interstitial or stromal tissue; L, lobule; M, mitotic figure; ME, gonadal mesentery / mesovary / mesorchium; O, oogonium; OL, ovarian lumen; PF, prefollicle cell; PO, pachytene oocyte; PS, pachytene spermatocyte; S, spermatogonium; SE, Sertoli cell; SO, somatic cell; SP, spermatocyst; SI, spermatocyte I; SII, spermatocyte II.

The ovarian cavity is formed completely by day 65, when numerous basophilic oocytes at the diplotene meiotic stage are also present. From days 80 to 100, the ovary consists mainly of follicles which contain characteristically large, perinucleolar oocytes with a basophilic cytoplasm: these previtellogenic oocytes are arrested at the diplotene stage of the first meiotic prophase and are characterised by the presence of multiple round nucleoli in the peripheral nucleoplasm. Each perinucleolar oocyte is covered by a continuous layer of flat follicle cells. Interstitial connective tissue is present both in the ovarian periphery and in the angular spaces between three or more follicles in the inner ovarian region.

In contrast to gonadal development in the female, signs of histological differentiation in the prospective males are not observed until day 72. Therefore, undifferentiated gonads on day 65 are certainly diagnosed as prospective testes. They retain the pearlike shape of undifferentiated gonads and are much smaller than the ovaries of females of the same age. Spermatogonia at this stage cannot be distinguished from the undifferentiated germ cells at day 38, and their most characteristic feature is a prominent nucleus with one or two nucleoli. By day 72, spermatogonial proliferation is occurring and blood vessels become evident in the dorsal region of the testis. On day 80, meiotic activity is apparent, and spermatocytes at early stages of meiotic prophase become numerous. Spermatocysts are formed by isogenic spermatocyte groups bound by the cytoplasmatic processes of Sertoli cells. Spermato-genesis progresses synchronously within each cyst. Spermatogonia are recognised as isolated cells, individually surrounded by the cytoplasm of a Sertoli cell. By day 100, the testis is divided into lobules composed by both Sertoli and germ cells, and which are separated by interstitial tissue. Spermatogenesis has reached the secondary spermatocyte stage.

Under laboratory conditions, mature gonads, i.e., ovaries containing full grown vitellogenic oocytes and testes containing mature spermatozoa can be observed in one year-old specimens, while the first spawnings are recorded towards month 16 after hatching.

The Adult Pituitary Gland

It is both an elongated and flat gland, in which the middorsal and caudal parts mostly consist of neurohypophyseal tissue. As it is found in other cichlids (e.g., Mattheij *et al.*, 1971) the infundibular recess of the third ventricle is large and extends as branches that penetrate the adenohypophyseal tissue. The adenohypophysis is divided into the rostral pars distalis, the proximal pars distalis and the pars intermedia, the endocrine cells of these areas showing different tinctorial and immunohistochemical properties. The adenohypophysis is penetrated by neurohypo-physeal infoldings, particularly in the pars intermedia.

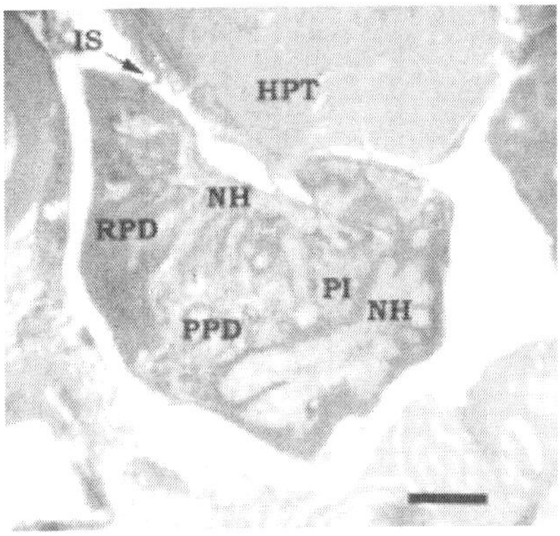

Figure: Sagittal section of the pituitary gland of an adult male Cichlasoma dimerus. Hematoxylin-eosin. The adenohypophyseal lobes (rostral -RPD- and proximal pars distalis -PPD-, and pars intermedia-PI) are indicated. The infundibular stalk (IS) is shown connecting the hypothalamus (HPT) to the pituitary (arrow). The neurohypophysis (NH) also shows its numerous infoldings into the pars intermedia, and in the rostral and proximal pars distalis too. Scale bar = 50 µm.

The distribution of endocrine cell types in the different adenohypophyseal lobes is summarised. Information about the antisera used in this study and the molecular weights of the detected *C. dimerus* pituitary hormones (Western blot analysis).

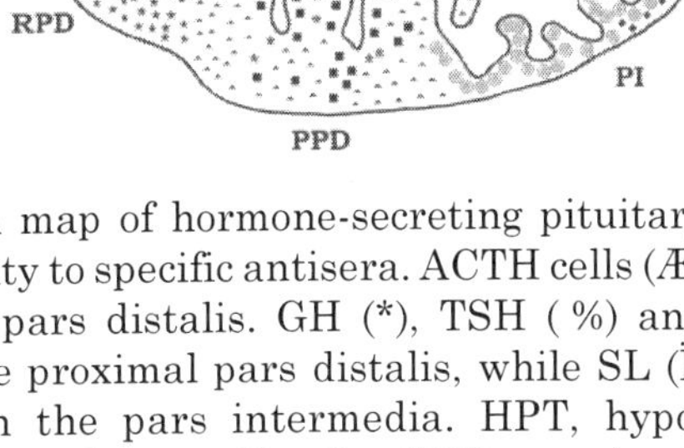

Figure: Distribution map of hormone-secreting pituitary cells, as shown by their immunoreactivity to specific antisera. ACTH cells (Æ%) and PRL (&) cells occur in the rostral pars distalis. GH (*), TSH (%) and GTH(²%) cells are mainly located in the proximal pars distalis, while SL (Ï%) and a-MSH cells (Æ%) are located in the pars intermedia. HPT, hypothalamus; PI, pars Intermedia; RPD, rostral pars distalis; PPD, proximal pars distalis; NH, neurohypophysis; IR, infundibular recess. Redrawn from Pandolfi et al. (2001a).

Rostral Pars Distalis

This lobe of the adenohypophysis is invaded by narrow neurohypophyseal prolongations (Pandolfi *et al.*, 2001a,b). In *C. dimerus,* two different cell types are typical of the rostral pars distalis, which are recognised by their immunoreactivity to antisera against prolactin and corticotropin (ir-PRL and ir-ACTH cells). Immunoreactive PRL cells are relatively large cells which form a compact layer in the periphery of the rostral pars distalis, while ir-ACTH cells are arranged in narrow bands or small follicles along the neurohypophyseal branches but without forming a continuous layer. Immunocytochemical labelling with the ACTH antibody was also observed in the pars intermedia, where the cells reacting with a-melanocyte stimulating hormone (ir-aMSH) cells are located. This cross-reactivity was likely due to the fact that both the corticotrops of the pars distalis and the melanotrops of the pars intermedia elaborate their distinctive hormones from the same large precursory molecule, namely propiomelanocortin (POMC) and that the whole a-MSH sequence is included in the ACTH mature molecule, so that the antisera against ACTH recognise a-MSH also (Eipper and Mains, 1981; Pandolfi *et al.*, 2003).

Proximal Pars Distalis

In this region only some broad neurohypophyseal branches are present but they do not reach the ventral part of the pituitary. Four different cellular types are distinguished immunochemically with heterologous antisera against growth hormone, follicle-stimulating hormone, luteinizing hormone and thyrotropin (designated as ir-GH, ir-FSH, ir-LH and ir-TSH cells). Immunoreactive GH cells form a layer surrounding the centrally located neurohypophysis.

Particularly, the identification of ir-FSH cells using heterologous antisera was hindered by their low cross-reactivity with the endogenous FSH. Antisera raised against bFSH and bLH from the mummichog, *Fundulus heteroclitus*, were successfully used to recognise gonadotrops in several teleost fishes, since they recognise conserved regions of FSH and LH in teleosts (Shimizu *et al.*, 2003a,b) and they were also successful in *C. dimerus* (Vissio *et al.*, 1996; Pandolfi *et al.*, 2001a). Gonadotrops are periodate reactive basophilic cells, which are localised mainly in the proximal pars distalis and the external border of the pars intermedia. Immunoreactive LH cells are located in central, ventral and marginal zones of the proximal pars distalis and in the external border of the pars intermedia. These cells are arranged in either cords or small groups along the whole proximal pars distalis and have conspicuous processes directed towards the neurohypophyseal infoldings and they showed a vacuolated cytoplasm (Pandolfi *et al.*, 2006). Immunoreactive FSH cells are detected in central and ventral zones of the proximal pars distalis and also in the external border of the pars intermedia. Immunoreactive FSH cells are either spherical or polyhedrical cells, with a vacuolated cytoplasm, and they appear either isolated or 2-3 cells groups. In general, they are smaller and less numerous than ir-LH cells, and are more anteriorly located in the proximal pars distalis. Our immunocytochemical analysis revealed that FSH and LH are mainly produced in different cell populations of the adult adenohypophysis in *C. dimerus*, as it was shown by both double-label immunohistochemistry and by the observation of adjacent sections (Pandolfi *et al.*, 2006). Finally, ir-TSH cells were seen forming discontinuous strands and small groups in the proximal pars distalis. However, their number was very small if compared with the other pituitary hormone cells of *C. dimerus*.

Pars Intermedia

The small lobules and cell rows of the most caudal part of the pituitary gland are separated by numerous irregular branches of neurohypophyseal tissue. In adjacent sections, somatolactin (SL) immunoreactivity was found in groups of periodate reactive cells which surround the most deeply located neurohypophyseal branches. In spite of the chemical similarities between GH, PRL and SL, no labelling with SL antisera was observed in either the rostral or the proximal pars distalis, where ir-GH and ir-PRL cells are located. Also, in adjacent sections, groups of periodate non-reactive cells can be detected in the pars intermedia, which corresponds to ir-aMSH cells.

Overview of Our Studies in the Adult Pituitary Gland

A clear-cut lobar distribution of the common vertebrate pituitary hormones was found, with a localisation of ir-PRL and ir-ACTH cells in the rostral pars distalis, the glycoprotein hormones ir-FSH, ir-LH and ir-TSH cells located in the proximal pars distalis (together with ir-GH cells), while the ir-SL cells (which are specific of fish) are located in the pars intermedia, where they are together with the non fish-specific ir-aMSH cells. In general, this lobar distribution would make the *C. dimerus* pituitary amenable to possibly interesting *in vitro* studies after microdissection and separate culture of the different lobes, since the chemically related PRL, GH and SL are produced in different lobes, as they are the chemically related ACTH and aMSH.

Cells involved in the production of the glycoprotein family of gonadotropins and thyrotropin are located in a single lobe (the proximal pars distalis) but they would still be amenable to experimental manipulation with different putative releasing hormones.

The Larval Pituitary

The cellular organisation of the pituitary gland during development in *C. dimerus* was described using immunocytochemical methods (Pandolfi *et al.*, 2001b; 2006). Although several cytochemical investigations of pituitary cells ontogeny have been reported in teleosts (Mal *et al.*, 1989; Cambre *et al.*, 1990; Nozaki *et al.*, 1990; Naito *et al.*, 1993; Saga *et al.*, 1993; Saga *et al.*, 1999; Villaplana *et al.*, 1997; Parhar *et al.*, 1998), ours are the only studies performed in cichlid fishes. This has the unfortunate derivation, however, of the difficulty of making inter-familial or even inter-ordinal comparisons in such a diverse class as the Teleostei. In the following paragraphs, developmental stages are indicated as either hours or days after hatching.

Newly Hatched Larvae (1 h)

They show an incipiently differentiated brain, which is curved over the ventral part of the body, and a large yolk sac. At this stage, the brain and the pituitary gland cannot be distinguished from each other and no immunoreactive cells were detected.

Vitelline Larvae (1.5 days)

The brain becomes straightened out and its ventricles are highlighted by the actively proliferating cells which surround them. Near the prechordal cartilage, which is in contact with the diencephalon, a mass of presumptive adenohypophyseal cells appears, but none of them is reactive to PRL, GH or SL antisera. At this stage ir-aMSH and ir-ACTH cells are detected in the posterior and anterior part of the developing pituitary, respectively. No immunoreactivity to either b-FSH, b-LH or b-TSH was detected.

Vitelline Larvae (2 days)

The pituitary gland is recognised as a rounded structure hanging from the floor of the diencephalon and which is formed by morphologically similar epithelial cells. The neurohypophysis has not yet been differentiated. A few small ir-SL cells are scattered in the caudal part of the gland, showing strong immunoreactivity. Immunoreactive PRL cells also appear at this stage in the rostral area of the adenohypophysis. Some few ir-GH cells are immunostained weakly, and are distributed in the central area of the developing pituitary gland. At this stage, ir-aMSH and ir-ACTH cells are found in the posterior and anterior part of the developing pituitary, respectively.

Opened-Mouth Larvae (4 days)

The pituitary gland is located over a ventral cartilage plate, and it still shows no clearly delimited lobes. The detected pituitary cell types (ir-SL, ir-PRL and ir-GH) increase both in their number and their immunostaining intensity. A compact group of ir-PRL cells begins to form at the more rostral part of the gland, but these cells are not arranged in follicles as in adult animals. Both ir-SL and ir-GH cells begin to increase in number and size but still without forming compact groups. From this day on, a topographic segregation of the three cellular types in a rostral-caudal direction is observed, which precedes the adult lobar distribution.

Free-Swimming Larva (6 days)

The infundibular recess develops at the junction of the diencephalon and the pituitary gland. The adenohy-pophysis becomes larger but the neurohypophysis is not yet formed. The segregation of the endocrine cellular types (ir-ACTH, ir- aMSH, ir-GH, ir-PRL and ir-SL cells) is more evident at this stage, approximating their final localisations; also, however, some few immunore-active PRL, GH and SL cells remain isolated in different regions.

Odd-Fins Larva (14 days)

The neurohypophysis becomes distinct in the caudal part of the gland but it is still a thin structure and does not show the adult deep infoldings. The adenohy-pophysis becomes even larger and elongated in shape with ir PRL and ir-ACTH cells occupying most of the rostral region (the prospective rostral pars distalis). The pars intermedia is distinct now, and is characterised by groups of ir-SL cells.

Prejuvenile Larva (21 days)

The first pituitary ir-FSH cells are detected in the anterior region of the adenohypophysis. From this stage on, the number of ir-FSH cells augments steadily. By day 90, when this ontogenetic study was ended, the number of FSH-ir cells was still increasing.

Prejuvenile Larva (26 days)

Two neurohypophyseal branches project into the adenohypophysis: a thicker one is directed towards the anterior adenohypophyseal regions while the other enters deep into the pars intermedia. At this time, a clear boundary between the pars distalis and the pars intermedia is seen, but there is still no clear demarcation between the rostral and the proximal pars distalis. Immunoreactive GH cells and SL-ir cells start to associate to the neurohypophyseal branches, both at the anterior adenohypophyseal regions and at the pars intermedia, respectively.

Juvenile (42 days)

The adult lobar topography is already distinct at this stage. Immunohistochemistry applied to adjacent sections showed that ir-SL cells are not periodate reactive, as they will be in the adult pituitary (Pandolfi *et al.*, 2001a). All the pituitary cell types, except ir-LH cells, are present at this stage. Close to the onset of sexual differentiation, which will occur by this time of development, the number of ir-FSH cells increases and they can be observed along the proximal pars distalis and the external border of the pars intermedia.

Juvenile (60 days)

Immunoreactive LH cells are first seen in central zones of the proximal pars distalis (two and a half weeks after the appearance of the first signs of ovarian differentiation).

Studies on Pituitary Ontogeny

Five immunoreactive cell types (PRL, GH, SL, ACTH, MSH) appear at very early stages of development, suggesting their functional importance for the fish larvae. Further studies focused on the ontogeny of hypothalamic factors that regulate their synthesis and secretion should still be performed. Gonadotrops become differentiated later: FSH cells appear prior to the onset of sexual differentiation, while LH cells appear during the sexual differentiation period.

Neuroendocrine Systems that Control Reproduction and Body Colour in C. Dimerus

Gonadotropin-Releasing Hormone (GnRH) Systems: The chemical structure of GnRH was initially discovered as a decapeptide which is present in the mammalian hypothalamus (Matsuo *et al.*, 1971; Burgus *et al.*, 1972) and is released in the portal system, and whose primary function was to induce pituitary gonadotropins release into the bloodstream. It was also found that it may act in neural tissue as a neuromodulator and/or neurotransmitter involved in eliciting reproductive behaviours (Ogawa *et al.*, 2006) and neuroendocrine reactions (Parhar, 1997). In addition to that, there is evidence suggesting that GnRH may play a role in the release of other pituitary hormones such as PRL, GH and SL (Weber *et al.*, 1997; Parhar, 1997; Stefano *et al.*, 1999; Vissio *et al.*, 1999).

Moreover, it was found that chemical forms other than the decapeptide also occurred in mammals and other vertebrates (Muske, 1993). At least three molecular GnRH forms exist in perciform species, which are identified after the species where they were originally found as sGnRH (salmon GnRH), sbGnRH (seabream GnRH) and cIIGnRH (chicken II GnRH). Their distribution patterns in the perciform central nervous system suggest the existence of three distinct neuronal populations located in (1) the olfactory bulbs, expressing sGnRH; (2) the preoptic area, predominantly expressing sbGnRH; and (3) the midbrain tegmentum, expressing cGnRH II. Based upon these observations and those of ontogenetic studies, it has been hypothesized that sGnRH neurons originate from the olfactory placode, cGnRH II neurons originate from a mesencephalic primordium, and two possible origins were proposed for sbGnRH neurons: the diencephalic floor and the olfactory placode (Parhar, 1997; Parhar *et al.*, 1998; González-Martínez *et al.*, 2002). This hypothetical triple origin of GnRH neuronal systems differs from that proposed for some fish and other vertebrates, according to which all forebrain and diencephalic GnRH neurons would originate from the olfactory placode (Schwanzel-Fukuda and Pfaff, 1989; Wray *et al.*, 1989; Muske, 1993; Schwanzel-Fukuda, 1999).

In a recent study (Pandolfi *et al.*, 2005), both antibodies against different perciform GnRH-associated peptides (GAPs), and riboprobes to different perciform GnRH and GAPs were used. These GAPs are useful markers of the different GnRH neuronal populations since they are produced from GnRH precursor molecules and they are colocalised within GnRH-expressing cells (Ronchi *et al.*, 1992; Polkowska and Przekop, 1993). Furthermore, markedly different peptides are associated with the different GnRH types. Therefore, these procedures result in a more specific marking than that obtained with antibodies and riboprobes to the smaller GnRH molecules themselves, both for immunohistochemistry and for *in situ* hybridization, respectively. Results showed that distribution of both sGnRH and sbGnRH neuronal populations in the *C. dimerus* forebrain are largely overlapped along the olfactory bulb, the nucleus olfacto-retinalis, the preoptic area and the ventral telencephalon. It was also shown (Pandolfi *et al.*, 2005) that sGAP immunoreactivity follows an antero-posterior gradient, with the most conspicuous immunoreactivity (both in terms of cell area and number) in the nucleus olfactoretinalis, and a lesser one in more posterior locations, as the preoptic area and the ventral telencephalon. In contrast, sbGAP cells show a postero-anterior gradient, with the most conspicuous immunoreactivity in the preoptic area and a weaker one in the nucleus olfactoretinalis and olfactory bulb.

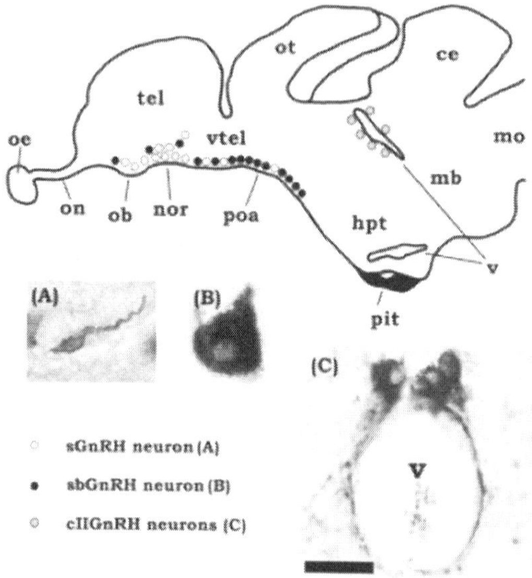

Figure: Distribution map of neurons producing the different types of GnRH found in the brain of Cichlasoma dimerus. The location of sGnRH, sbGnRH and cIIGnRH neurons was studied by immunoreactivity of their respective GnRH-associated peptides (GAPs). Insets A, B and C show details of sGnRH, sbGnRH, and cIIGnRH neurons at the same magnification; scale bar = 20 μm. Encephalic regions and brain nuclei are indicated as: ce, cerebellum; hpt, hypothalamus; mo, medulla oblongata; mb, midbrain; nor, nucleus olfacto retinalis; ob, olfactory bulb; oe, olfactory epithelium; on, olfactory nerve; ot, optic tectum; pit, pituitary; poa, preoptic area, tel, telencephalon; v, ventricle; vtel, ventral telencephalon. Redrawn from Pandolfi et al. (2005).

Overview of Our Studies on Neuroendocrine Systems that Control Reproduction (GnRH)

Our results showed that: (1) GnRH neuronal populations in the forebrain (sGAP and sbGAP) show an overlapping pattern, (2) sGAP projections are mainly located in the forebrain and contribute to the pituitary innervation, with cGAPII projections being mainly distributed along the mid and hindbrain and not contributing to pituitary innervation, whereas sbGAP projections are restricted to the ventral forebrain, being the most important molecular form in relation to pituitary innervation; (3) sbGAP neurons have an olfactory origin; (4) GAP antibodies and GAP riboprobes are valuable tools for the study of various GnRH systems, by avoiding the cross-reactivity problems that occur when using GnRH antibodies and GnRH riboprobes alone.

MCH and a-MSH Systems

Melanin-concentrating hormone (MCH) is a 17 amino acids cyclic peptide synthesized in the hypothalamus, which together with a-melanocyte-stimulating hormone (a-MSH) regulates the aggregation and dispersion of chromatophores in teleosts (Fujii, 1993). Other factors like norepinephrine, adenosine triphosphate (ATP), PRL and SL have also been reported as potential regulators of teleost skin colour changes.

In adult *C. dimerus*, ir-MCH neurons are mainly located in anterior and midhypothalamic positions (Pandolfi *et al.*, 2003) as they are also found in other teleost fishes (Baker and Bird, 2002). Two kind of ir-MCH cells are found in *C. dimerus*: magnocellular neurons located in the nucleus lateralis tuberis, and parvi-cellular neurons, which are less numerous ir-MCH cells, localised between the third ventricle and the lateral ventricular recess. Magnocellular neurons send their projections through the anterior and posterior infundibular stalk to the three adenohypophyseal lobes, but mainly to the pars intermedia. Other axons are also projected to the thalamus, preoptic area and telencephalon. Another group of small MCH-ir neurons located outside the nucleus lateralis tuberis show delicate immunoreactive nerve fibres that do not contribute to pituitary innervation.

The distribution of ir-MCH magnocellular neurons and their fibres suggests two principal roles in *C. dimerus*, (1) as a neurohypophyseal hormone, and (2) as a hypothalamic regulator of pituitary hormones. The presence of ir-MCH fibres in the thalamus, preoptic area, ventral and dorsal telencephalon also suggests a potential role of MCH as a neuromodulator/neurotransmitter, as proposed for amphibians, sauropsids and mammals (Bittencourt *et al.*, 1992; Cardot *et al.*, 1994; Vallarino *et al.*, 1989). The role of ir-MCH cell bodies located outside the nucleus lateralis tuberis, as well as the nerve fibres found in different brain regions is less clear, but they might be involved in the control of feeding, as it has been proposed for mammals.

In *C. dimerus* larvae, ir-MCH cells appear 4 days after hatching, in a proliferating zone located on the floor of the anterior hypothalamus (Pandolfi *et al.*, 2003). At this stage, the MCH innervation can be detected as a dense bundle of fibres penetrating into the developing pituitary. The number of immunoreactive neurons and fibres increase with age and by day 30 (prejuvenile stage) cell bodies and fibres already show an adult arrangement in the nucleus lateralis tuberis. By day 6 post-hatching, another group of small ir-MCH neurons develops in the nucleus periventricularis posterior, located close to the third ventricle in the medial part of the hypothalamus. These cells have few and thin projections that do not seem to reach the pituitary. The number of these small ir-MCH neurons decrease with age and by day 42 they are not detectable anymore. In *Sparus auratus*, Mancera and Fernández Llebrez (1995) ob-tained similar results. Probably, these neurons differentiate in the ependymal layer and then migrate to the nucleus lateralis tuberis.

Involvement of SL in Background Colour Adaptation

SL is a pituitary hormone present at least in Actinopterygian fish and in the Sarcopterygian lung-fish, *Protopterus annectens*. This hormone is expressed in the pars intermedia and belongs to the GH/PRL family. A single main function of SL remains elusive, but it is thought to be involved in different regulatory processes, such as those controlling the stress response, circulating calcium and phosphate levels, fat metabolism, as well as some reproductive processes (Kawauchi and Sower, 2006). Furthermore, studies suggest that SL is involved in background colour adaptation (Kawauchi and Sower, 2006). In *C. dimerus*, a melanin dispersing effect is observed in fish exposed during ten days to a black background, while a concentrating one is observed in those exposed to a white background. When we compared Western blot-detected pituitary SL content in fish adapted to either black or white backgrounds, we observed an important increase in those adapted to the black background. In addition, fish exposed to a black background had more SL-ir cells with a greater ir-area. On the other hand, fish chronically adapted to a black background showed a larger number of ir-aMSH cells, while fish adapted to a white background had more numerous ir-MCH neurons in the nucleus lateralis tuberis (Cánepa *et al.*, 2006). Taken together, these results strongly suggest that SL interacts with both aMSH and MCH in the regulation of background colour adaptation in *C. dimerus*.

Regulation of SL Release by GnRH and MCH

As we have seen, many studies have focused on SL function, but there is scarce information on the regulation of its secretion. In cultured pituitaries of the rainbow trout, *Oncorhynchus mykiss*, Kakizawa *et al.* (1997) have shown that sGnRH, corticotropin releasing factor and serotonin stimulate SL release. Also, in the masu salmon, *Oncorhynchus masou*, Onuma *et al.* (2005) observed that the addition of sGnRH to primary pituitary cell cultures elevated SL mRNA expression.

In *C. dimerus*, we have shown by double label immunofluorescence that ir-SL cells are in close contact with both ir-MCH and ir-GnRH fibres in the pars intermedia (Cánepa *et al.*, 2008). We then performed *in vitro* pituitary cultures of intact in order to investigate a possible releasing or inhibiting function of MCH or GnRH on SL cells. We found that MCH stimulated SL release from pituitaries of male *C. dimerus* in a dose dependent manner; but no such an effect was observed in females. On the other hand, incubation with GnRH showed similar results than the MCH assay in male pituitaries. Although female pituitaries did not evidence significant differences between treatments, we observed a noticeable increasing tendency. In conclusion, our results show that SL, like other pituitary hormones, would be under hypothalamic control (MCH and GnRH) and that they are implicated in diverse physiological processes including background adaptation and reproduction (Cánepa *et al.*, 2008).

The Roles of Pituitary and Brain Gonadotropins.

The gonadotropins follicle-stimulating hormone (FSH) and luteinizing hormone (LH) are glycoprotein hormones produced in the proximal pars distalis of *C.dimerus*. In fishes, as in other vertebrates, these hormones are considered of central importance in the control of gametogenesis and gonadal steroidogenesis (Blázquez *et al.*, 1998). In several tetrapod and teleost species, they are formed by two protein chains, namely the a and b subunits (Yoshiura *et al.*, 1999; Yaron *et al.*, 2001). The a-FSH and a-LH subunits have identical aminoacid sequences in a given species, while b subunits have different sequences, and are recognised by different specific receptors, and are thus responsible for their different actions.

In salmonid species characterised for spawning only once a year, the different roles of FSH and LH have been studied in detail (Suzuki *et al.*, 1988; Tyler *et al.*, 1991). The functions of FSH, as well as the variations of its circulating levels during the reproductive cycle, are not completely understood in many non salmonid species examined so far, though FSH is generally considered as a vitellogenic hormone, while LH is generally related to gonadal maturation and steroidogenesis.

Several developmental studies conducted in teleost fishes, showed varying plasma levels and different expression patterns of FSH and LH in the pituitary (Dufour *et al.*, 2000). Taken together, these results suggest specific functions for both FSH and LH at the onset of sexual maturation and in the control of gametogenesis. As we have mentioned above, the identification of FSH cells using heterologous antisera was very difficult to perform due to cross-reactions or low specificity between FSH molecules from different species (Vissio *et al.*, 1996; Pandolfi *et al.*, 2001a). Antisera raised against the mummichog *Fundulus heteroclitus* FSH and LH were successfully used to recognise GtH cells in several teleost fishes, because they recognised conserved regions of FSH and LH in teleosts.

The localisation, characterisation and ontogeny of gonadotrops in the pituitary of *Cichlasoma dimerus* have been already addressed this review. In this species, pituitary ir-FSH cells appear before the onset of sexual differentiation, while ir-LH cells differ-entiate several days thereafter. This variation in the timing of immunoreactive FSH and LH expression was also observed in several teleost species (Mal *et al.*, 1989; Nozaki *et al.*, 1990; Saga *et al.*, 1999; Magliulo-Cepriano *et al.*, 1994). The results of these studies suggested that in salmonid species and in the platyfish (Schreibman *et al.*, 1982; Parhar *et al.* 1995), as well as in *C. dimerus*, the differentiation of ir-FSH cells precedes that of ir-LH cells. In the pejerrey *Odontesthes bonariensis*, however, ir-LH cells are known to appear before ir-FSH cells.

In salmonids, FSH, but not LH, is involved in the onset of gonadal growth and development (Saga *et al.*, 1993). Also, FSH was shown to be involved in the onset of meiosis in an amphibian (Ito and Abe, 1999). Since ir-FSH was expressed in pituitary cells of sexually undifferentiated prejuvenile larvae of *C. dimerus* (30 days after hatching), we decided to study the effects of this hormone on larval gonadal explants. LH effects was not studied since ir-LH cells do not appear in the pituitary until day 60, i.e. two and a half weeks after the first indications of gonadal differentiation appear. When the undifferentiated gonadal explants were left for 15 days with no hormone in the control culture medium, no sexual differentiation occurred, while exposure to human recombinant FSH for the same period resulted in 80% of cases differentiating as prospective ovaries (N=80) (Pandolfi *et al.*, 2006). These results reinforced our previous hypothesis about a temporal correlation between the innervation of pituitary cells by sbGnRH fibres and the subsequent release of FSH to circulation, and the onset of gonadal sex differentiation in both sexes.

In the course of our anatomical study of the *C. dimerus* pituitary (Pandolfi *et al.*, 2001a; 2006), ir-FSH and ir-LH fibres were detected in both the neurohypo-physis and the infundibular recess. These fibres were deriving from cell bodies in the parvicelullar and magnocellular preoptic nuclei, and in the hypothalamic nucleus lateralis tuberis. Also, the developmental appearance of both brain- and pituitary-derived gonadot-ropins was studied showing that ir-LH neurons appeared on day 15 after hatching, while ir-FSH pituitary cells and neurons appeared on day 21), and ir-LH pituitary cells appeared only on day 60.

To corroborate these findings, specific PCR products of b-LH and b-FSH were amplified from cDNA obtained from pituitaries and the preoptic area and hypothalamus of adult *C. dimerus* during the reproductive season. The partial nucleotide sequences of mRNA from *C. dimerus* brain and from the preoptic-hypothalamic brain fragment consisted of approximately 258 bp for b-LH and 276 bp for b-FSH (Pandolfi *et al.*, 2009).

Furthermore, we settled an individual pituitary *in vitro* culture system to study the possible modulatory effect of brain derived gonadotropins on pituitary hormone secretion. Whole pituitary explants were cultured with different concentrations of *Fundulus heteroclitus* LH or FSH and the culture media were analyzed by Western blot. Exogenous LH produced a dose-dependent increase in pituitary b-LH and SL release and an increase in b-FSH release. No effect was observed on both GH and PRL release. Exogenous FSH produced an inhibition in b-LH release, a dose-dependent increase in b-FSH and SL release, and no effect on PRL and GH release. This is the first work showing an effect of brain-derived gonadotropins on different vertebrate pituitary cells (Pandolfi *et al.*, 2009).

Remarks and Future Studies

Cichlasoma dimerus is a very interesting fish species for the study of reproductive biology because of several characteristics already discussed in this review: (1) it is easy to maintain and breed under laboratory conditions; (2) it has a moderate size which allows extraction of tissue samples and pituitary dissection easily and fast; (3) it shows a very interesting reproductive behavior; (4) it is possible to specifically identify their pituitary endocrine cells and hormone-producing neuronal populations; (5) its brain anatomy is well known; and (6) most of the important reproductive events (sexual differentiation and maturity, annual reproductive cycle) have been already described.

Future research in this species will be aimed to understand how the brain influences behavior, and how behavior correlates with the expression of brain hormones. To analyze these interactions we will study reproduction, perhaps the most important process in an animal's life. All the presented data about neuroanatomy, neuroendocrinology, reproduction and development of this species will be extremely useful for us and any other research group interested in using *C. dimerus* as a vertebrate species which displays social interactions that would be manageable in the laboratory.

After several years of observations we realised that this species has a very sophisticated social system with features we can manipulate in the laboratory. Our current experiments are aimed at modifying the social circumstances, to analyze the consequences of our manipulations on behavior and on the previously characterised neuroendocrine systems. This species is also easy to observe in the field, so a lot of behavioral experiments could be performed not only for answering ecological and behavioral questions but also to address evolutionary issues (they have incredibly robust social interactions that are clearly important for natural selection).

The accessibility of its habitat enables observations in nature, something that is not as easy to achieve in "model species" such as medaka or zebrafish. For all these reasons and all the data gathered in this review we think that *C. dimerus* might turn into a valuable vertebrate model in the future.

Attempt to Improve the Reproductive Efficiency of Nile Tilapia Brood Stock Fish

Tilapia aquaculture is and will continue to be an important fish, particularly for the lesser-developed countries in the tropics. Nile tilapia (*Oreochromis niloticus)* are considered as the most common and popular fish in Egypt. Egypt, a country where, arguably, the farming of tilapia has its roots, where tilapia culture is believed to have originated some 4000 years ago. Tilapia consist 36% of the Egyptian production from fish culture and occupy the 10th order concerning the world production from aquaculture. Hence, Egypt produces 20% of the world tilapia capture and 12% of the world farmed tilapia. The culture of *O. niloticus* in Egypt has recently developed into a major industry. This industry, however, is still growing in a remarkable way with apparent intend towards intensification that pressurizing the need of enormous number of seeds. Many limitations associated tilapia fry production under the prevailing

Egyptian conditions were described by El-Gamal. Also, brood stock husbandry and spawning techniques are constantly upgraded as Egyptian hatcheries require a high number of good quality eggs to satisfy the needs for aquaculture, so rigorous management of large numbers of brood stock are necessary for mass production of eggs and fry due to the complex reproductive biology and asynchronously spawning with relatively small number of eggs produced per spawning.

Accordingly, the development of more elaborated forms of brood stock management is crucial to improve fry yield and system efficiency. Today, it is widely accepted that effective seed production demands a thorough understanding of the special husbandry and particular nutritional requirements of brood stock fish which significantly affect fecundity, survival, egg size and egg and larval quality. The objective of the present research was to study the possibility of improving reproductive performance of Nile tilapia fish using some feed additives.

Chapter 2

Materials and Methods

The present study was carried out during Nile tilapia hatching season of 2008 (June and July) in two phases, the first was to study the effects of three commercial feed additives on females (Therigon®, Nuvisol Hatch P® and Gibberllic acid) and fourth one (L - carnitine) on males brood stock, concerning their gonads characteristics and some reproductive traits. The second phase was to select treatments of both 0.5 g Therigon® and 2 g Nuvisol Hatch P® / Kg diet (for females) and 700 mg L - carnitine / Kg diet (for males) to apply in a mating trial to be evaluated via hatchability.

Experimental Management of the 1st Phase

A field study was conducted in a private earthen pond fish farm located at Alabhar belonging to Alhamol, Kafr Alshiekh governorate. Fourteen Habas (each 3 m width × 6 m length × 0.5 m water depth and 1 m total depth) were constructed in a two Feddans earthen pond. The first ten Habas were stocked with females brood stock of one year (yearlings) Nile tilapia fish (average body weight of 150 g) from the same farm. The other four Habas were stocked with males brood stock of (the same age as the females, but of an average body weight of 200 g) Nile tilapia fish from the same farm also. Each Haba was stocked with twenty fish. This study was a feeding trial to test the effects of some commercial dietary supplements on Nile tilapia (*O. niloticus*) propagation. The experimental feeding began on the 20th June till the 8th of July, where the feed was offered to fish twice daily, at a daily feeding rate of 3% of the fish biomass in each Haba. The feed additives were purchased from the local market and added directly to a mash diet, which was purchased also from the local market (contained 90.31% dry matter, 80.88 % organic matter, 23.81% crude protein, 5.47% ether extract, and 9.43% ash) after the proximate analysis according to AOAC and moistened to be pelleted via a hand mincer. The feeding continued for males and females before mating. Water of fish rearing in each individual Haba was tested daily for some water quality parameters including water temperature, pH value, and dissolved oxygen concentration.

Dietary Treatments During the 1st phase: Fish were fed on a basal ration (BR) with or without (control) the tested feed additives which were:

1- Therigon® powder for veterinary use, manufactured by Adwia Co., S. A. E. 10th of Ramadan city, Egypt. Each 1 g contains Alpha - Amino - p - hydroxyhydrocynnamic acid, 1000 g package as GnRH stimulant (Batch No. 0601116).
2- Nuvisol Hatch P®, imported by Khirat Alnile Co., 27 Alferdos Buildings, Flat 43, Nasr city, Egypt from Newtrix Co., Belgium, in 500 g package. Each 1 Kg contains the following vitamins (in mg): B_1 4000, B_2 5000, B_3 4000, B_6 2000, B_9 1000, B_{12} 20, PP 10000, Biotin 50, and L - carnitine 30000.
3- Gibberllic acid (C_{19} H_{22} O_6), type analysis, Art. 3930, M. W. 346.38, M. P. 225 °C, GA_3 content 90 %, 1 g package, Batch No. 43124, imported from Lobal Chemie, Pvt. LTD, 2042 Bombay, India.
4- L - carnitine powder from Mebaco, Egypt.

Criteria Measured at the end of the 1st Phase

After the 19 days feeding trial of the separate sexes of brood stock, three fish from each experimental Haba were catched to collect blood, seeds (eggs), and milt for different measurements and determinations. Tri sodium citrate was used as an anticoagulant for blood collection. Blood determinations for follicle stimulating hormone (FSH), luteinizing hormone (LH) and progesterone hormone were done using commercial colorimetric kits (Diamond, Diagnostic, Egypt), and milt analyses (count, motility, forward, sluggish and dead percents) were done too.

Experimental Management During the 2nd Phase

Six Habas (similar to those used in phase one, in the same earthen pond, at the same private farm) were used in the second phase (beginning from the 9th of July) of this study.

Statistical Analysis

Data collected were statistically analyzed using SAS, when ANOVA-test was significant ($P \leq 0.05$), least significant difference was calculated to differentiate between means.

Results

There were no significant differences among all Habas concerning the tested water quality parameters. Therefore, the data were presented as overall means ± standard errors (SE).

The tested parameters showed very suitable water conditions for rearing tilapia.

Table: Females' sexual parameters of brood stock Nile tilapia as affected by the dietary supplementations for 19 days feeding trial in Habas in an earthen pond (means* ± SE).

Haba No.	*Fish weight, g*	*Ovaries weight, g*	*Egg number / fish*	*Egg No./Kg BW***	*Egg diameter, mm*	*GSI****
1	178.1[b] ± 17.1	8.45[ab] ± 1.35	769.3[b] ± 34.4	4341[a] ± 224.8	1.55[a] ± 0.12	4.71[a] ± 0.31
2	211.7[b] ± 5.70	9.00[ab] ± 0.90	689.0[bc] ± 99.0	3267[bcd] ± 491.1	1.55[a] ± 0.12	4.26[a] ± 0.76
3	227.0[b] ± 22.1	9.25[ab] ± 1.75	926.0[a] ± 9.00	4114[ab] ± 360.1	1.55[a] ± 0.12	4.03[a] ± 0.38
4	227.0[b] ± 32.1	11.5[a] ± 0.05	698.3[bc] ± 14.3	3037[cde] ± 286.8	1.67[a] ± 0.00	5.14[a] ± 0.70
5	224.5[b] ± 3.50	9.00[ab] ± 1.20	935.5[a] ± 85.2	4171[ab] ± 445.0	1.25[b] ± 0.00	4.02[a] ± 0.59
6	239.0[ab] ± 19.1	10.7[ab] ± 1.75	519.0[de] ± 74.2	2161[ef] ± 138.2	0.96[c] ± 0.04	4.42[a] ± 0.38
7	223.2[b] ± 20.3	7.5[ab] ± 2.50	537.5[d] ± 12.6	2423[def] ± 163.7	1.67[a] ± 0.00	3.49[a] ± 1.44
8	288.9[a] ± 16.9	7.90[ab] ± 0.40	630.0[bcd] ± 14.1	2191[ef] ± 176.8	1.67[a] ± 0.00	3.28[a] ± 0.83
9	178.5[b] ± 4.96	5.95[b] ± 2.35	602.0[cd] ± 0.00	3376[bc] ± 93.9	1.67[a] ± 0.00	3.30[a] ± 1.22
10	223.9[b] ± 15.9	10.1[ab] ± 0.05	391.5[e] ± 12.6	1754[f] ± 68.5	1.67[a] ± 0.00	4.41[a] ± 0.40

* Means with the same letter within the same column don't differ significantly (P e" 0.05).

**: BW=body weight.

***: GSI = gonado - somatic index = gonads weight (g) × 100 / fish weight (g).

Tables present data of males reproductive traits tested including testes weight and GSI as well as milt quality parameters. Although there were no significant differences among treatments and the control; yet, the control was more pronounced in fish weight, testes weight, GSI, and sperms count than the treatments. But the motility and dead percentages were better in fish group of Haba No. 12 (700 mg L - carnitine / Kg diet) followed by Haba No. 13 (900 mg L - carnitine / Kg diet) concerning motility, forward, sluggish and dead percentages.

Table: Males' gonado - somatic index of brood stock Nile tilapia as affected by the dietary supplementations for 19 days feeding trial in Habas in an earthen pond (means* ± SE).

Haba No.	*Fish weight, g*	*Testes weight, g*	*GSI***
11	283.05 ± 0.85	4.050 ± 0.05	1.425 ± 0.01
12	236.10 ± 22.3	2.350 ± 0.95	0.965 ± 0.32
13	250.00 ± 33.0	2.550 ±0.05	1.035 ± 0.12
14	263.55 ± 14.0	2.450 ± 0.75	0.915 ± 0.24

*:Means don't differ significantly (P ³ 0.05).

**: GSI = gonado - somatic index = gonads weight (g) × 100 / fish weight (g).

Table 6: Data of some quality parameters of milt (using Hämocytometer slide) collected from the tested males brood stock Nile tilapia as affected by the dietary supplementations for 19 days feeding trial in Habas in an earthen pond.

Dead, %	***Sluggish, %***	***Forward, %***	***Motility (viability), %***	***Count, × 106***	***Haba No.***
35	45	20	55	660	11
25	40	35	75	450	12
30	40	50	60	350	13
30	45	35	45	390	14

Females fish of the 3rd and 5th Habas reflected lower concentrations of (FSH) but higher concentrations of either (LH) or progesterone hormone, comparing with the other Habas' fish. This is in good accordance with the fecundity which was given in Table as shown. Also, male fish of the 12th and 13th Habas had sera with higher levels of FSH, LH, and testosterone comparing with the other treatments. This also confirms the previous results of the milt analysis for its quality parameters.

Hydroxybutyrate, could be biosynthesized in plant and animal cells via lysine, methionine, and some vitamins like B_6, C, nicotinic acid and folate. It is an essential cofactor of fatty acid metabolism (it provides energy by transporting long and medium chain fatty acids to mitochondria to act as fuel). Deficiency in carnitine is associated with male infertility. Since L-carnitine provides an energetic substrate for the spermatozoa in the epidydimis, contributes directly to sperm motility and may be involved in the successful maturation of the sperm. Carnitine lowers triglycerides and raises the high density lipoprotein levels. L-isomer of carnitine is more effective than DL -

isomer. Cellular parameters of the seminogram have been previously shown to correlate with L-carnitine concentration in the seminal fluid. Carnitine is involved in maintaining an active oxidative phosphorylation (OXPHOS). Therefore, it was strongly suggested that relationship between carnitine secretions, seminal quality and OXPHOS activities could be because of a parallel response to the same regulatory event. L-carnitine improved semen quality and histological characteristics of the testes. Generally, a low level of L-carnitine enrichment provides several protective effects in fish reared under intensive pond-culture conditions.

Pituitary homogenate induced artificially maturing and increased both serum testosterone and estradiol. Gonadotropin-releasing hormones (GnRHs) bind to the specific receptor on the gonadotrophs to activate the synthesis and release of gonadotropins (follicle stimulating hormone or FSH and luteinizing hormone or LH), which in turn control gonadal maturation, gametogenesis and gamete release. Not only do teleosts exhibit the highest variety of GnRH variants, but recent data and whole genome analyses indicate that they may also possess multiple GnRH receptor. Synthetic analog of gonadotropin-releasing hormone was used for inducing ovulation and enhancing spermiation in brood fish. Moreover, territorial males of African cichlid, *Haplochromis burtoni*, characterised by aggressive and reproductive activity, have significantly larger hypothalamic gonadotropin-releasing hormone (GnRH)-containing neurons and larger testes than nonterritorial males.

Dietary inclusion of 1 g Nuvisol Hatch P® / Kg diet and 1 g Therigon® / kg diet before mating realised good females reproduction performance. Also, dietary supplementation with 700 and 900 mg L-carnitine / Kg diet before mating gave better results of males' reproductive performance. But, because of the high feed cost due to the additives cost, 0.5 g Therigon® / kg diet as pretreatment for @& only (3rd Haba of the 2nd phase, mating), 2 g Nuvisol Hatch P® / Kg diet as pretreatment for @& only (5th Haba of the 2nd phase, mating), followed by 0.5 g Therigon® and 700 mg L-carnitine / Kg diet for @& and B&, respectively (4th Haba of the 2nd phase, mating), respectively were the best economically.

It could be recommended to use such commercial feed additives for improving reproductive performance of Nile tilapia brood stocks to offer enough seeds for fish farms .It is recommend also to make other trials on different other additives at economical levels.

Antibiotic Use in Food Animals

Broad concern and confusion about antibiotic use on farms have been expressed by the general public and policymakers, according to a white paper published from the 2011 National Institute for Animal Agriculture Symposium Proceedings on Antibiotic Use in Food Animals: A Dialogue for a Common Purpose. Key reasons for this concern and confusion can be traced to numerous factors including:

- Antibiotic use in food animals is not a black-and-white issue. It is a complex issue that is all too frequently over simplified by both critics and proponents.
- Misunderstanding that a concern is not equivalent to risk.
- The disconnect between consumers and agriculture (and those in agriculture), with most consumers being at least three generations removed from the farm.
- Activist messaging, the media and the Internet are often inaccurate and misleading regarding antibiotic use, and in particular antibiotic resistance and its relationship to use, in food-animal production.

Antibiotic use in food animals is highly regulated by the U.S. Food and Drug Administration Centre for Veterinary Medicine (FDA CVM). Regulatory oversight provides assurance in the development of safe products. Education and training encourages producers to have a strong relationship with their veterinarians and provides guidance on the responsible use of antibiotics. Government surveillance and testing ensures that no harmful residues, as established by the FDA, enter the food supply. The use of antibiotics in food-animal production has elicited concern about antimicrobial resistance (AMR). The development and dissemination of antibiotic resistance is a highly complex, multi-faceted issue.

The issue of AMR is not strictly an issue of science-based decision-making. Like many other aspects of food production, the issue of AMR inherently invokes differing opinions, and, given the intricacies and size of the modern food production system, it's a foregone conclusion that any policy issue will invoke many different perspectives. The symposium was intentionally designed to be respectful of all opinions, including the varying politics and values related to modern food production practices. Therefore, it is essential to initiate broad dialogue around the role of antibiotics in animal agriculture and the issue of antibiotic resistance.

Ultimately, key questions concerning antimicrobial use in human and animal populations revolve around:

- Who should benefit and by how much?
- What is "over" or "unacceptable" use?
- What is acceptable risk?

Given these questions, concern about AMR bacteria is not equivalent to risk. Risk to humans exists only if there is a causal pathway from the AMR bacteria in food-producing animals and humans. Risk is not a result of the hazard alone, but also in conjunction with the exposure (or dose). Estimated farm-to-fork risk from on-farm antibiotic use is extremely low. In fact, the alternative risk of suboptimal animal health may be higher than the risk of on-farm antibiotic use.

Although activists often claim that antibiotics used in livestock and poultry are the cause of human antibiotic resistance, the attributable risk of human disease outcomes, e.g., additional illness days due to resistance, associated with the use of antibiotics in food-producing animals is unknown, and studies show it is extremely low.

Livestock-associated methicillin-resistant Staphylococcus aureus (LA-MRSA) is a new and potential health concern for persons in the swine/livestock industries, but current evidence indicates it is not a concern for the broader community as it has not been shown to be a significant cause of the overall MRSA burden in humans. Since LA-MRSA is not a food-borne pathogen, it is not a food safety or public health concern.

Common on-farm practices that consumers should know about regarding food-animal production include:

1. Modern livestock farms increasingly involve licensed veterinarians who advise on health management decisions.
2. Vaccines are used to protect animals from various illnesses.
3. Sick animals are treated with medicines, such as antibiotics, to restore their health, and protections are in place to ensure that their meat or milk is safe for people.
4. The FDA approves the use of all new animal drugs after testing and confirming animal safety and human food safety.
5. If antibiotics are administered to cure a sick animal, the animal itself — in the case of meat production — or animal products — such as milk — are not allowed to enter the food supply until the withdrawal period has passed and the medicine has

sufficiently cleared the animal's system. The required periods for withdrawing medication are specific for each drug and species and are approved by the FDA based on research studies of residues in edible tissues.

Animal agriculture stands on the belief that healthy animals make safe food. Those who produce food animals agree with the veterinarian's oath: "Being admitted to the profession of veterinary medicine, I solemnly swear to use my scientific knowledge and skills for the benefit of society through the protection of animal health and welfare, the prevention and relief of animal suffering, the conservation of animal resources, the promotion of public health, and the advancement of medical knowledge." Today, however, consumer concerns may cause two phrases in the oath to collide: "prevention and relief of animal suffering," and "promotion of public health."

Individuals dedicated to food-animal production recognise that proper production practices, such as maintaining flock and herd health by preventing disease through biosecurity and the use of vaccines, help minimise the need for antimicrobial use. Medically important antibiotics used in food-animal agriculture should be used under the supervision of a veterinarian to combat disease on a case by case basis. Access to antibiotics to prevent and treat animal disease is essential for the health of food animals as well as to ensure a safe food supply.

It is critical that policy actions regarding the use, limited use, or non-use of antibiotics in food-animal production be based on science with a commitment to carefully weighing the benefits and costs as well as the desired — and potentially undesired — consequences of any specific course of action.

In final analysis, the ultimate priority about antibiotic use going forward is the development of well-established, science-based criterion in the regulatory decision-making process. Simultaneously, the livestock industry should remain focused on continual improvement of good animal husbandry practices and disease prevention.

There should be an achievable, unified goal of "One Health: Healthy People, Healthy Animals, Healthy Food."

Probiotics in Aquaculture: Do They Work?

There is growing interest in the use of beneficial bacteria, probiotics, as an alternative strategy to antimicrobial compounds for disease prevention and control in aquaculture. These naturally occurring bacteria exert their beneficial effects on the host by modifying the microbial community associated with the host, by ensuring improved use of the feed or enhancing its nutritional value, or by enhancing the host response towards disease.

Establishing a healthy gut is the key to animal health and performance. Probiotic solutions are a sustainable alternative to antimicrobial compounds, as the latter have led to the development of resistance among bacterial populations. Great care must be taken in the selection of probiotic strains, since it is essential to ensure that bacteria are beneficial to the host and able to thrive in the gut and aquatic environment.

Lactic acid producing bacteria, e. g. Enterococcus faecium, have been the focus of much interest. Combined with other beneficial bacteria, multi-species products have proven that it is possible to provide synergistic bacteria with complementary modes of action to enhance health and zootechnical performances of cultured aquatic species. But why is there still doubt?

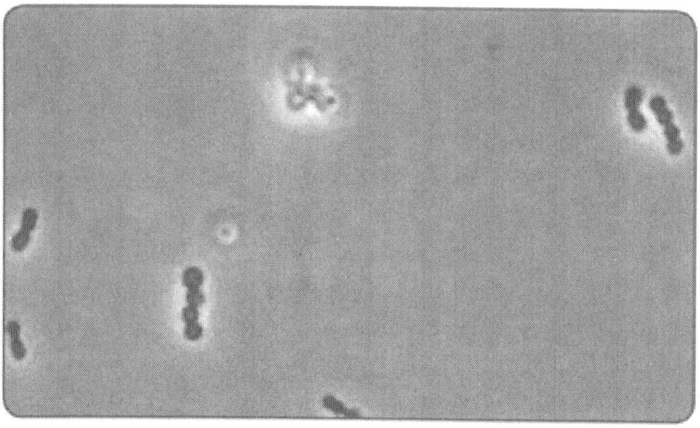

Figure: Microscopic picture (1000-fold magnification) of Enterococcus faecium

Probiotics and Their benefits

There is growing interest in the use of beneficial bacteria, probiotics, as an alternative strategy to antimicrobial compounds for disease prevention and control in aquaculture. These naturally occurring bacteria exert their beneficial effects on the host by modifying the microbial community associated with the host, by ensuring improved use of the feed or enhancing its nutritional value, or by enhancing the host response towards disease. At the animal level, probiotics improve the growth and survival of fish and shrimp by modifying the host-associated or ambient microbial community.

Feed Application of Probiotics for Intestinal Health

A well-established intestinal microflora is crucial for the growth and health of the animal, since the microflora has impacts on:

- nutrition,
- the prevention of pathogenic infections,
- the integrity and function of digestive organs,
- the development of the immune system

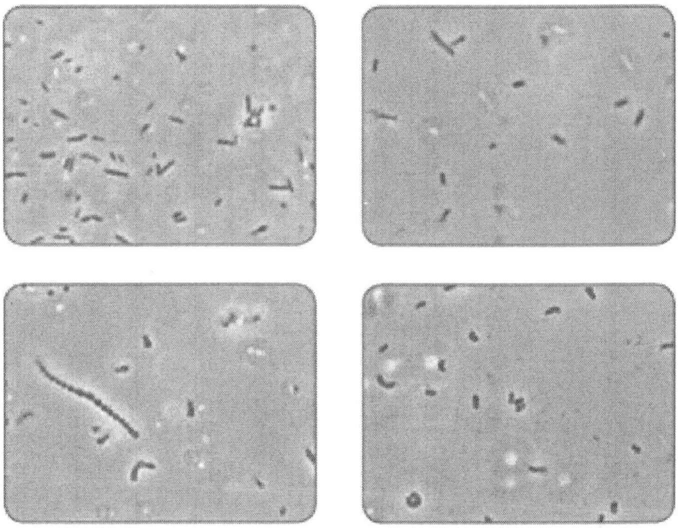

Figure: Microscopic pictures (1000-fold magnification) of Vibrio harveyi, Yersinia ruckeri, Streptococcus agalactiae and Aeromons veronii

Thus, the management of the gut flora is important for the ability to prevent infections with enteric pathogens and to guarantee a well-functioning and effective digestion of nutrients that result in good growth performance parameters.

Probiotic bacteria modulate the gut microflora towards a favourable composition. Hence, selection criteria of probiotics for aquaculture should be based on their antagonism towards pathogens (through competitive exclusion), their growth, attachment to intestinal mucus and production of beneficial compounds (Vine et al., 2004). Since aquatic species are cultured under different conditions salinity and optimum temperature range should also be considered for selection of the right probiotic strain. Another important factor is that probiotics lack pathogenicity. After all, probiotics should contribute to efficient production in a sustainable way, promoting healthy and robust animals (Brittain et al., 2002).

Lactic acid bacteria (LAB) are potential probiotic candidates in aquaculture and are also known to be present in the intestine of healthy fish (Balcázar et al., 2008). Enterococcus faecium is one of the most commonly used lactic acid producing bacteria in animal nutrition and has become a focus of attention for use in commercially farmed aquatic species. Enterococci are Gram-positive, facultative anaerobic bacteria which are widely distributed in nature.

The use of the bacteriocin-producing E. faecium with probiotic properties alone or in combination with other beneficial intestinal bacteria was investigated in different studies.

In vitro studies using the agar spot method (Rosskopf, 2010) have shown that Enterococcus faecium (strain IMB 52, BIOMIN) has inhibition properties against a wide spectrum of aquatic pathogens including Yersinia ruckeri, Vibrio harveyi, Streptococcus agalactiae and Aeromonas veronii. Similar observation was made by Swain et al. (2009) who proved the inhibitory activity of E. faecium isolated from brackishwater fish against V. harveyi and V. parahaemolyticus. This demonstrates the potential applications of E. faecium from fish intestine for controlling pathogenic vibriosis in shrimp culture.

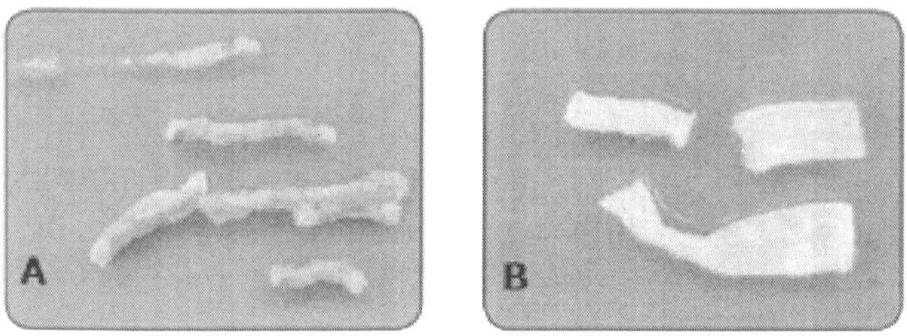

Figure: Faecal materials were removed from the intestine; B) Intestinal epithelium without faecal material

Panigrahi et al. (2007) examined immune modulation including cytokine gene expressions of rainbow trout (Oncorhynchus mykiss) and demonstrated that these parameters were improved by probiotic feeding of freeze-dried Lactobacillus rhamnosus, Enterococcus faecium or Bacillus subtilis (109 CFU/g) after 45 days. Particularly the fish fed the E. faecium strain showed better performance which could possibly be linked to the suitable ambient temperature conditions of this strain. Temperature is a major environmental factor controlling microbial growth and the ideal conditions differ among microorganisms. E. faecium was found to be more psychrotolerant than the other two bacteria, growing well at temperatures ranging from 12 to 30 °C. Rosskopf (2010) showed that the optimum temperature range for E.

faecium (strain IMB 52, BIOMIN) reaches as far as to 37 °C. In vitro studies were conducted by the Faculty of Natural Resources, Prince of Songkla University, Thailand, to demonstrate that E. faecium (strain IMB 52, BIOMIN) is also able to populate the intestine of Nile tilapia (Oreochromis niloticus) and induce a positive impact on bacterial ecology of the gut by inhibiting Vibrio spp. through competitive exclusion. E. faecium was even detected in the fish gut and faeces 10 days after product administration.

Why is There Still Doubt?

Although there is experimental evidence that the prophylactic use of beneficial bacteria can improve health and performance of cultured aquatic species, there are still some doubts. Results can be affected by improper management methods and product quality, such as:

- incorrect application methods (e. g. simultaneous use of chemicals or antibiotics)
- incorrect claims of probiotics, which can´t be fulfilled (e. g. for white spot syndrome)
- too low bacterial concentrations (colony forming units should be above 108/g)
- poor bacterial stability during production and storage

Figure: Backyard production of "probiotics"

High levels of viable organisms and stability during production and storage are important criteria for the selection of suitable strains. The safety of strains must be carefully assessed, as well, and transmission of antibiotic resistance or virulent plasmids must not take place. Of further great importance are the survival and growth of beneficial bacteria in culture conditions, and their ability to colonize the gut of the aquatic animals. Additionally to the improper management of probiotics, the use of "fake" probiotics, which have unconventional manufacturing processes and/or an undefined blend of bacteria, cultivated in the "backyard" (Picture 4), can lead to doubts about significant effects and benefits of probiotics.

Continuous research has been undertaken to develop new products for modern and sustainable aquaculture. For the production of a probiotic product in conventional production facilities, the seed should come from a respective cell master bank stored at -80 °C which serves as a standardized source as inoculum for a working cell bank. The entire fermentation process has to be performed under fully sterile conditions. Picture 5 shows the major production steps of single fermentation processes, from the bacteria seed to the dried lyophilisate. AquaStar®, BIOMIN′s probiotic product line for aquaculture, is a well-defined, multi-strain probiotic product, consisting of a blend of single-strain fermented bacterial strains.

There is plenty of evidence that probiotics supplemented in feed and water are effective in aquaculture application. However, the success of probiotic supplementations depends on the strains, concentrations and management used. Effective probiotics are based on selected strains and controlled production conditions. Poor results are many times related to low-quality and inadequate probiotics.

It has been shown that well-defined probiotic strains, e. g. Enterococcus faecium, are able to reduce pathogenic bacteria, thus improving gut health, and enhancing performance and efficiency in production of aquatic species.

New Technologies in Tuna Breeding

Unlike smaller species, broodstock management for large pelagic finfish such as tuna involves many difficulties, said Christopher Bridges, Heinrich-Heine Universität, at Aquaculture Europe 2011. Charlotte Johnston, TheFishSite editor reports.

The handling of such large fish of over 60 to 80 kg such as the bluefin tuna precludes many of the techniques used previously for smaller fish such as sea bream and sea bass.

New technologies are available through the EU programmes, including environmental monitoring of both cage and fish temperatures to predict spawning, broodstock identification, sex and tagging together with gamete induction and parentage of offspring and traceability.

Environmental and Fish Monitoring

This task is accomplished by using cage based light and temperature data loggers (HOBO) which could be sampled underwater using a submersible shuttle system. Fish data-loggers (Star-oddi) were mounted in syntactic foam floats and were implanted under water using a speargun. Both types of loggers could be individually programmed to record at set rates and times.

The results of the cage data-loggers were compatible with the recordings made from the Fish data-loggers. Temperature within the cage could be monitored quite easily and predictions made as to when the water temperature would reach 24°C.

Broodstock Identification, Sex and Tagging

This was done using a novel system developed to both tag and take a muscle biopsy at the same time. Fish were DNA-fingerprinted using molecular techniques and at the same time biomarkers (Vtg and ZRP) were used to determine sex. Tags and biopsy sampling is possible using both jab stick and speargun applications.

The broodstock could easily be tagged with ID markers and PIT tags and from the muscle biopsy both sex and DNA fin-printing was carried out enabling an approximate 90 per cent certainty in sex discrimination. A speargun based system for small muscle biopsies required for broodstock fin printing has been developed recently by the HHU Biology Workshop.

For the induction of spawning the tags used have been developed over a number of years moving from plastic carriers for hormones to titanium mounts together with detachable suture material. Application has been by jab stick and speargun and if required the direct implantation of custom-made or hormone pellets can be carried out.

Gamete Induction

Using a series of pre-prepared colour-coded tags together with hormonal implant carriers individual fish could be tagged and hormonally induced or treated underwater. This could involve either GnRHa to induce spawning in mature fish to estradiol implants to improve gamete production.

Parental contribution to gamete production i.e egg production is possible using either microsatellite markers or mitochondrial DNA which give differing resolutions of parental contribution.

The use of these new novel methods in aquaculture allows a more complete control of broodstock management and monitoring, concluded Mr Bridges. For open water gamete production the monitoring of temperature and light conditions are of great importance.

Since the size of the fish precludes the normal practices of removal from water then semi non-invasive techniques such as tagging, muscle sampling and hormonal induction can be carried out successfully underwater.

If the goal of sustainable aquaculture is the continued supply of viable gametes from the broodstock then the use of these new technologies will allow the control and performance monitoring of individual broodstock and further the ultimate goal of traceability from parent to egg to the table can be reached for the consumer.

Assessing Pollution Effects in Aquatic Environments

Many estuarine and coastal aquatic environments have been sinks for industrial and agricultural effluents for hundreds, perhaps thousands of years. Michael Leaver and Stephen George from the University of Stirling, look at what effects this may have had on aquatic environments.

There are now estimated to be over 100,000 distinct chemical compounds marketed in the European Union alone, and knowledge of the environmental fates and concentrations is restricted to only a small proportion known to be particularly harmful and persistent.

Against this background there has been concern that long-term exposure to complex cocktails of chemical contaminants may be affecting stocks of fish and other animals which live in estuarine and coastal areas. In fact surveys of flounder have shown higher incidences of tumours and fin rot in some polluted estuaries in the UK, other parts of the North Sea and in North America.

Although this sounds alarming, it has been difficult to provide the evidence that pollution is the cause of such ill-effects, or that it is the cause of any declines in population numbers. This is because we know very little about the biological effects of complex mixtures of chemicals, and estuaries, particularly, are highly dynamic and highly exploited environments in which population numbers fluctuate for many reasons not related to pollution.

However, over recent years it has become apparent that particular types of chemicals can have dramatic and measurable effects on individuals. For example, the effect of chemicals which mimic the natural hormone, estradiol, can induce a protein, vitellogenin, many hundred-fold levels above normal in wild, exposed fish. Similarly PCBs can induce a protein, CYP1A, to very high levels.

These proteins and their genes have been proposed as 'biomarkers' of pollution, meaning that their measurement in wild flounder, or another animal, can provide evidence that those individuals have been are derived from a point source, such as a sewage outfall, or in a spill situation, such an oil spill. In situations where fish are exposed long-term to complex mixtures of chemicals, biomarkers have proved less effective.

This may not be so surprising because most biomarkers are discovered by acute, short-term, laboratory exposures to single chemicals, a situation quite unlike most environmental exposure scenarios. Very recently techniques for measuring the levels of many thousands of genes simultaneously in practically any organism have been developed. We have applied this technology in an experimental situation to discover what effects longterm exposure to real polluted estuarine sediments have in flounder. Funded by NERC and in collaboration with Marine Scotland, Aberdeen, we established mesocosms, large tanks, containing sediments collected from polluted UK estuaries and from an unpolluted estuary.

We also collected live flounder from the unpolluted estuary and then put them into the mesocosms for seven months. Thus, the only difference between the mesocosm systems was due to the pollutant status of the sediments. All other natural variables found in estuaries, such as genetic background, salinity, temperature, food source etc were eliminated.

Following this exposure we measured the expression of thousands of genes in these flounders. When polluted flounder were compared to clean it was clear that none of the commonly used biomarker genes were affected, but that the expression of groups of genes with functions in the immune system and in apoptosis was increased.

Apoptosis is a cellular process which brings about the self-destruction of a cell following either an extrinsic signal, or the intrinsic sensing of unsustainable cellular damage. The genes we have found point to the involvement of intracellular chemical damage, which in turn induces the intrinsic apoptotic pathway. Thus, this study points to a mechanism by which long-term exposure to complex chemical mixtures can adversely affect fish, and also provides new candidate biomarkers for these situations.

The Nutrition and Feeding of Farmed Fish and Shrimp

By the Food and Agriculture Organisation of the United Nations - The classification, structure and biological function of the essential nutrients are described; including proteins and amino acids, lipids, carbohydrates, vitamins, minerals, and energy. The quantitative dietary nutrient requirements of fish and shrimp are critically reviewed, and suggestions made for dietary nutrient levels within practical complete diets for carnivorous and omnivorous fish and shrimp species.

Atlantic Cod, Genes and Plant Proteins

New studies of cod genes show that as much as 75 per cent of marine proteins can be replaced by plant proteins in feed for farmed Atlantic cod without triggering stress genes.

On a global basis, fish meal resources are limited and researchers are looking for other sources of protein for use in fish feed. Plant meal is an alternative source of protein that is already widely used. Many fish feed trials have been conducted in order to establish how the farmed Atlantic salmon utilise nutrients in plant meal, but fewer studies have been carried out on farmed Atlantic cod.

Previous studies carried out by the National Institute of Nutrition and Seafood Research (NIFES) indicate that cod can consume a feed consisting of up to 75 per cent plant protein without any significant reduction in the growth. Nevertheless, new plant ingredients may cause more subtle health effects in the fish.

Therefore, we wanted to investigate whether a high level of plant protein in the feed would affect cod health, or have any other subtle negative effects on the Atlantic cod. We therefore examined the stress gene expression. The results of our studies showed that the stress genes appeared to be unaffected, says researcher Kai Lie from the Aquaculture Nutrition Research Programme at NIFES.

Why Examine Stress Genes?

Stress is a broad term. For instance, fish may experience stress when they are given a new feed containing a high level of plant proteins instead of fish meal. By studying the stress response at gene level it is possible to identify harmful effects of plant meal in the feed given to cod at an earlier stage than can be revealed by looking at parameters such as growth and survival, says Mr Lie.

As the liver plays an important role in protein metabolism, the cod liver was examined to see if there was any evidence of changes in the genes involved in stress and the metabolism of proteins.

Liver samples were taken from cod that had been given feed containing 75 per cent plant protein. After testing the expression of around 500 genes we found no changes to indicate a stress response in the cod. The liver cells were apparently unaffected by the high level of plant meal, says Mr Lie.

We also examined genes involved in protein turnover (synthesis and metabolism) and the gene expression was lower for a few of these genes in the cod that had been fed with 75 per cent plant protein. This shows that the cod was affected to some extent, even though the growth was normal. Whether this has any significance for growth and protein metabolism in a prolonged life periodis a matter that requires further study, says Mr Lie.

About the Cod in This Study

Cod weighing around 1.7 kg were given a feed containing either 25, 50, 75 or 100 per cent plant proteins. The remainder of the protein level in the feed was fish meal. The control group was given a feed containing only fish meal as a source of protein. Mr Lie chose to focus on the cod group that had been given 75 per cent plant protein as they did not show reduced growth. The cod that were given 100 per cent plant protein showed reduced growth and were therefore not suitable for more in depth studies to establish whether cod react to stress before the stress reduces their growth.

The feeding experiment lasted for 28 weeks. Protein consists of amino acids. The composition of amino acids in the feed met the requirements set by the National Research Council, Canada, but the balance of amino acids was different between the fish meal and the plant proteins. The cod that were not given this feed were compared with cod that had been given 100 per cent fish meal as the source of protein.

Vertebrate Vitellogenin Gene Duplication in Relation to the 3R Hypothesis

The spiny ray-finned teleost fishes (Acanthomorpha) are the most successful group of vertebrates in terms of species diversity. Their meteoric radiation and speciation in the oceans during the late Cretaceous and Eocene epoch is unprecedented in vertebrate history, occurring in one third of the time for similar diversity to appear in the birds and mammals. The success of marine teleosts is even more remarkable considering their long freshwater ancestry, since it implies solving major physiological challenges when freely broadcasting their eggs in the hyper-osmotic conditions of seawater. Most extant marine teleosts spawn highly hydrated pelagic eggs, due to differential proteolysis of vitellogenin (Vtg)-derived yolk proteins. The maturational degradation of Vtg involves depolymerization of mainly the lipovitellin heavy chain (LvH) of one form of Vtg to generate a large pool of free amino acids (FAA 150–200 m*M*). This organic osmolyte pool drives hydration of the ooctye while still protected within the maternal ovary. In the present contribution, we have used Bayesian analysis to examine the evolution of vertebrate Vtg genes in relation to the "3R hypothesis" of whole genome duplication (WGD) and the functional end points of LvH degradation during oocyte maturation. We find that teleost Vtgs have experienced a post-R3 lineage-specific gene duplication to form paralogous clusters that correlate to the pelagic and benthic character of the eggs. Neo-functionalization allowed one paralogue to be proteolyzed to FAA driving hydration of the maturing oocytes, which pre-adapts them to the marine environment and causes them to float. The timing of these events matches the appearance of the Acanthomorpha in the fossil record. We discuss the significance of these adaptations in relation to ancestral physiological features, and 1propose that the neo-functionalization of duplicated Vtg genes was a key event in the evolution and success of the teleosts in the oceanic environment.

Ancestral Gene Landscapes

Gene duplication has long been recognised as a powerful mechanism in evolution. As noted by Haldane, and Ohno the products of such events may acquire new functions (neo-functionalization), survive due to dosage effects (sub-functionalization), become dysfunctional due to mutational insertions, deletions and/or substitutions (pseudogenes), or be lost, without obligate deleterious effects to the host. Whole genome duplications (WGD) have been argued to be the bases of major events in the evolution of vertebrates. However, the massive gene loss that may occur in the aftermath of WGD, can mask the ancestral ploidy of an organism. Similarly the propensity for certain lineages to acquire transposable elements; to independently duplicate genes through cis- or trans- events; to retain or lose extra chromosomes (aneuploidy) or a complete set of chromosomes (tetraploidy, haploidy, diploidization) during meiosis or early germline mitosis with subsequent deletion or translocation of genes or chromosomes, further complicates the deciphering of ancestral gene landscapes. Indeed, while teleosts are considered to have doubled their genes in relation to humans, chromosomal loss, or diploidization seems to have been rampant in this group, since similar numbers exist between teleosts (~20–25) and rodents and primates (19–22 autosomes). As a result, the number of WGD, and whether such events have occurred at all, has remained controversial.

Recent comparative genomic analyses of paralogous chromosomal regions (paralogons) between humans and other vertebrate and invertebrate model organisms have provided convincing evidence that two rounds of WGD occurred during early chordate evolution. Similar analyses between humans and the gene maps of actinopterygian fishes revealed a double conserved synteny between the chromosomes of the teleosts compared to those of a human, and thus demonstrated that a fish-specific WGD first postulated by Ohno and later proposed by Amores *et al.* had indeed occurred. Most recently, Crow *et al.* provided strong evidence that the fish-specific WGD took place after the separation of the crown group of teleosts from bowfin, a neopterygian fish. Hence three rounds (R1, R2, R3) of WGD have occurred during the evolution of teleost fishes from early chordate roots, and traces of these ancient events should be detectable in the many thousands of extant species. Any phylogenetic reconstruction of genes should thus reconcile the chronology of these events with the proposed tree model and the fossil record.

A Brief History of Teleosts

The origin of the Actinopterygii dates back to paleonisciform fishes during the late Silurian (410 mya) some 40 million years after separating from the Sarcopterygii. One of the most primitive actinopterygian fishes was *Cheirolepis*, a freshwater fish that did not enter seawater. While some descendents, such as *Mimia* and *Bigeria* did apparently enter seawater, their lines became extinct long ago and thus did not contribute to the teleost lineage. Only those groups that remained in freshwater exist today. These are the Chondrostei including the Polypteriformes (bichirs) and Acipenseriformes (sturgeons and paddlefishes), and the neopterygian Semionotiformes (gars) and Amiiformes (bowfin). Despite anadromic behaviour in sturgeons, all of these fishes are obligate freshwater spawners, as indeed are the more ancestral Hyperoartia (lampreys). The monophyletic primitive teleosts, did not appear in the fossil record until 235 mya, but are suggested to have arisen 250–290 mya. Despite some teleost fossils being recorded from marine deposits, it is not known whether this is the result of anadromous behaviour, or represents true marine species spawning in the ocean. The oldest modern teleosts, however, are the

Osteoglossomorpha (bony tongues), all of which, both extinct and extant, were and are freshwater species. The first extant modern marine teleosts appeared as Elopomorpha (e.g. eels and tarpons) during the Jurassic while subsequent Clupeomorpha (e.g. herrings and shads) had both freshwater and anadromous forms. The most successful group of freshwater teleosts, the Ostariophysi (carps and catfishes), were the next to appear (~245 mya), and today represent more than 75% of all freshwater fishes of the world with ~8000 species (30% of all teleosts). However, they only achieved their great diversity in freshwater long after the appearance of the Acanthomorpha. It was not until the appearance of Acanthomorpha ~55–125 mya that the unprecedented radiation and speciation of the teleosts occurred. Extant Acanthomorpha comprise ~16,000 species, equivalent to 86% of non-ostariophysan teleosts. According to Maissey:

> *"Charts showing the range of fishes through geological time suddenly sprout hundreds of new families. Many of the fossils come from the marine strata of Monte Bolca in Northern Italy. These and other fossils from around the world leave no doubt that modern family-level acanthomorph diversity is rooted in the early Eocene (about 55 million yeas ago), and that this teleost explosion was the most dramatic evolutionary radiation ever seen in vertebrate history, eclipsing the evolution of mammals and birds in numbers of families and species."*

The basis of this teleost explosion has never been satisfactorily explained. Although the long freshwater ancestry of the teleosts is documented in the fossil record, it can also be inferred from their physiology. All teleosts are hypo-osmotic to seawater, and hyper-osmotic to freshwater. The hypo-osmotic condition of marine teleosts as a group is unique among animal taxa, and is generally assumed to reflect their freshwater past. A related argument for a freshwater origin concerns the presence of a glomerular kidney and a distal tubule segment, the main functions of which are to produce copious volumes of hypo-osmotic urine to counteract the osmotic influx of water in the freshwater teleost. Glomerular kidneys are also present in the more ancient Chondrichthyes, with glomerular filtration rates (GFR: 1–4 mL $\cdot$ kg^{-1} $\cdot$ h^{-1}) equivalent to the GFRs of freshwater teleosts (2–10 mL $\cdot$ kg^{-1} $\cdot$ h^{-1}). Indeed the presence of glomerular kidneys led Homer Smith to suggest that the Chondrichthyes, and even the extinct placoderms, had a freshwater origin. He argued that the glomerulus, the sole function of which is to filter extracellular fluids, could only have arisen in the presence of excess water – i.e. freshwater. In contrast, the marine teleosts have considerably reduced glomeruli, and fractional GFRs (0.1–0.5 mL • kg^{-1} • h^{-1}) compared to their freshwater relatives and the Chondrichthyes. Some marine teleosts are even aglomerular (a unique condition among vertebrata), relying entirely on secretion rather than filtration for ion regulation. Reduction or loss of glomeruli may be considered an adaptation facilitating the conservation of water in the functionally arid marine environment.

Spawning in the Sea – the Water Problem

The long freshwater ancestry of teleosts implies that also their eggs, which are freely broadcast into the environment by their oviparous parents, had become adapted to the freshwater conditions in the rivers and lakes. When, however, the teleosts eventually started to spawn in the sea, the eggs met new, opposite osmotic problems. The yolk osmolarity is similar to the parental body fluids, ~350 m*Osm*, and thus hypo-osmotic to seawater. Hence, instead of an osmotic influx, the problem in seawater is a continuous osmotic water efflux. Thus, since the osmoregulatory systems (mitochondrial rich cells, intestine, kidneys, and gills) are not yet developed, the egg must be endowed with a water reservoir before it can be spawned in the hyper-osmotic seawater. A water reservoir in the egg is a prerequisite and a necessary pre-adaptation before the teleosts could complete their life cycles in the oceans and truly establish themselves as marine organisms. Without this egg feature, the adult fish had to remain anadromous and return to their freshwater habitats to spawn their eggs. Indeed anadromy is typical of the more primitive fishes including lampreys, sturgeons, shads and salmonids.

Already Fulton and Milroy noted the remarkable volume increase during oocyte maturation in the eggs of marine teleosts, and intuitively proposed that an osmotic mechanism caused a watery fluid to be secreted into the oocytes. This unique oocyte hydration in marine teleosts has been confirmed for a wide range of species with the degree of hydration being much greater in species that spawn pelagic eggs (pelagophils) compared to those that spawn benthic eggs (benthophils). In benthophils the mechanism mostly involves the differential movement of inorganic ions, and a high concentration of the amino acid analogue taurine, while in pelagophils, free amino acids (FAA) due to the maturational proteolysis of vitellogenin (Vtg)-derived yolk proteins, are the main osmolytes driving the hydration. This growing body of literature shows that up to three forms of Vtg are expressed and differentially processed in pelagophil and benthophil teleosts.

Vitellogenins and the Organic Osmolyte Pool of Pelagic Eggs

In teleosts Vtg genes are linearly organised as large monomeric structures with multiple sub-domains consisting of a lipovitellin heavy chain (LvH), phosvitin (Pv), lipovitellin light chain (LvL), and a von Willebrand factor type D domain (vWFD) that is split into a beta-component (ß2) and a C-terminal coding region (CT). Once assembled and secreted by the liver, Vtgs are taken up via clathrin-mediated endocytosis by the growing oocytes and cleaved by cathepsin D (CatD) in the early endosomes to form the primary yolk proteins. The CatD processing represents the primary cleavage event in the degradation of yolk proteins. Recent studies have shown that developmentally regulated V-class ATPases (proton pumps) acidify the yolk platelets during oocyte maturation. This disassembles the crystalline, or in some species non-crystalline yolk and activates other cathepsins (CatL or CatB, dependent upon species) that hydrolytically attack the yolk proteins. This latter processing is known as the secondary cleavage event and is unique to marine teleosts. In some freshwater species, such as zebrafish, electrophoretic band-shifts of the LvH occurs during oocyte maturation due to nicking, but no proteolysis, and no buildup of FAA occurs (Finn, unpublished data), which would be osmotically disadvantageous to the freshwater embryo. During oocyte maturation in marine benthophils the yolk proteins are either not processed, or are partially cleaved and hydrolyzed with the release of a small pool of FAA, while in marine pelagophils, the yolk proteins are not only cleaved but undergo extensive differential proteolysis of particularly one of the LvH domains resulting in the buildup of the large pool of FAA. The transient increase in osmolarity during oocyte maturation causes water influx via specialised aquaporins that are temporally inserted in the plasma membrane during this period. This hydration provides the early embryo of marine teleosts with a vital water reservoir before a drinking mechanism is developed.

Fulton also noted that the greater hydration of the pelagic eggs compared to the benthic eggs caused them to float and hence acquire their pelagic nature. More recently, in a comprehensive review of the early life history stages of fishes and their characters, Kendall *et al.* revealed that most extant marine fishes, regardless of systematic affinities, demersal or pelagic habits, coastal or oceanic distribution, tropical or boreal ranges, spawn pelagic eggs. These findings suggest that the rise of the pelagic egg was an important event in the evolution of the teleosts.

Vitellogenin Genealogy

Since no evidence of the ancestral spawning habits of teleost fishes has been recorded in the fossil record, we can only infer it. The best means of achieving this is through Bayesian phylogenetic inference for genes involved in the reproductive physiology of the parent and the survival of the embryo. The hill-climbing, and proportional hill-hopping Markov Chain Monte Carlo algorithm of Bayesian methods is superior to traditional methods of phylogentic inference. Since Vtg-derived yolk proteins are the major components of the egg that sustain the embryo during early development, and are implicated in the pre-adaptation to the marine environment, we analyzed all available vertebrate Vtg genes (sequenced and genomic) and the related homologue apolipoprotein B (apoB) in relation to WGD. Previous studies have shown that apoB, which is related to Vtgs, is also incorporated in the yolk of birds, and may represent a neo-functional product of gene duplication. We therefore included the first ~1000 amino acids of apoB proteins, which represent the large lipid transfer module (LLTM) that is homologous the LvH of Vtgs. Other studies have claimed or continue to cite that the phosvitinless class of Vtg, which lacks the polyserine domain, is more closely related to insect Vtgs. We therefore included cnidarian and molluscan Vtgs, which predate the arthropods, as outgroups in order to reconcile the topology of the tree with the fossil record, currently accepted phylogenies, and the "3R hypothesis".

The notion that R3, which is estimated to have occurred prior to the appearance of the crown group of teleosts >290 mya , , is the primary cause of the diversity of teleosts seems improbable due to the ~200 million year gap between the WGD and the rapid diversification. Based upon the present analyses, we propose that the radiation and speciation of the acanthomorph teleosts is rooted in the adaptation of their eggs to the marine environment. We argue that the origin of the LLTM, which predates the bilateria, is the molecular harbinger of key adaptations that facilitated the dramatic radiation, while retention of 3R gene products latently contributed to speciation of the Acanthomorpha during the late Cretaceous and Eocene epochs.

Sequences

Based on our recent sequencing of full length Vtg genes from haddock and Atlantic halibut (Finn, unpublished data), related members of this family of low density lipoproteins were identified using the NCBI BLAST interfaces. Homologies were confirmed using the blast 2 sequence tool. Using this approach we identified 38 full length vertebrate Vtg homologues, and 56 partial sequences of which only 10 salmonid sequences were included due to their sufficient length and position at the N-terminal domain. We further accessed all available Vtg constructs for vertebrates from the Ensembl genome database. This resulted in the inclusion of 8 more constructs from chicken, zebrafish, 3-spined stickleback, medaka and torafugu. In addition we included the first ~1000 amino acids of apoB proteins from 5 model organisms. In total ~83, 000 aa were aligned in 73 sequences.

Since there is some debate in the literature as to the antiquity of the phosvitinless Vtg genes, two invertebrates (galaxy coral and Pacific oyster) were used as outgroups.

In order to corroborate these analyses and confirm the number of gene forms within teleosts, we retrieved all available Vtg genes from currently sequenced teleost genomes. This resulted in separate analyses of 2 genes from torafugu, 11 genes from 3-spined stickleback, 6 genes from medaka, and more than 16 genes from zebrafish. Due to recent updates of the zebrafish genome, we included both the new fragments from Ensembl release 41 (October 2006), and full genes from release 38 (April, 2006) to match the published data of Wang *et al.*.

Multiple Sequence Alignments

Multiple amino acid (aa) alignments of the Vtg homologues were achieved using several programs. Initial alignments were performed on the full length Vtg protein data set with default settings using t-coffee, Clustal W, Muscle and Probcons. Extensive modelling was performed using Blossum and PAM matrices and by varying gap-open and gap-extension penalties. This strategy allowed us to make a high quality alignment for the first ~1075 aa, and the last ~500 aa. The remaining polyserine segment was separately aligned using t-coffee and re-inserted into the full alignment. In addition, we further modelled the alignment in Clustal X profile mode using the lamprey structure mask for Vtg. This latter approach allowed us to minimise gaps in regions associated with secondary structures. Based on these outputs we manually adjusted the sequences to give a final full length multiple alignment. A second data set included the 10 partial salmonid sequences that were added to the full data set using Clustal X in profile mode. All data sets were then converted to codon alignments using the

University of Bergen computational biology unit alignment to coding tool. To determine the most likely tree topology, the full aa and codon alignments were analyzed using phylogenetic programs, and then re-examined after removal of the signal peptides, and polyserine domain which showed the least consensus. The alignments were further examined after removal of regions that contained gaps in more than 70% of the taxa, and again after removal the C-terminal vWFD domain that only occurs in most VtgA type homologues. Two separate sets of the first 270 aa and 810 nucleic acids, which included the partial salmonid sequences, with and without gaps were also analysed. A summary of the domains analysed is shown below the multiple sequence alignment.

Phylogenetic Analyses

Bayesian analyses (MrBayes 3.1.2;) was used for each of the aligned aa and codon data sets. The following settings were used for codon alignments: nucmodel = 4by4, nst = 2, rates = gamma; and aa alignments: aamodel = mixed, with 1,000,000 generations, sampled every 100 generations using 4 chains and a burning of 3,500. For each run, a Majority rule consensus tree together with posterior probabilities from the last 6,500 trees, representing 650,000 generations, was rendered with ATV. The codon alignments were examined for clock-like behaviour using the MrBayes strict-clock model. Speciation or duplication events were inferred using the method of Zmasek and Eddy. To corroborate the Bayesian results, maximum likelihood analyses of the codon alignments and maximum parsimony and neighbour-joining analyses of the aa alignments with 1000 bootstrap replicates were conducted using PAUP 4.0b10. In order to understand the evolution of the sub-domain structure of Vtgs, the ratio of non-synonymous (Ka) to synonymous (Ks) nucleotide substitution rates were estimated using the Ka/Ks web service at the computational biology unit, University of Bergen.

Evolution of Vertebrate Vitellogenin Genes

As in previous studies we found good homology between the LLTMs of vertebrate LvHs and apoBs, and for the separate vWFD domains. Computed Ka/Ks ratios for each of these domains over the whole tree were not significantly different (0.411±0.181 and 0.375±0.177 between the LvH and vWFD domains, respectively), but were significantly lower ($p<0.05$) than for the LvL domains (0.492±0.163). Excluding the Pv domain, these Ka/Ks ratios indicate that the LvL domain is the least conserved, and that vertebrate Vtgs are not under strong purifying selection, but are evolving. This notion is corroborated by the Bayesian strict-clock model analyses, which showed that the evolution of the Vtg genes is not clock-like.

Although apoB is known to be related to Vtgs, the N-terminal LLTM has yet to be examined in the context of WGD. To investigate whether apoB could have arisen as a consequence of R1, we accessed the NCBI and Ensembl databases for related vertebrate and invertebrate lipoproteins. 128 of these sequences were used in an alignment consisting of ~100 000 aa for Bayesian analysis. We followed the methodology as described above. These studies revealed that apoB is not a directly duplicated product of Vtg, but is secondarily derived from microsomal triglyceride transfer protein. To illustrate this, we included a kink in the descending branch of the tree that leads to the apoB cluster.

Consensus Bayesian reconstruction of the evolution of vertebrate Vtg proteins was well supported by codon analyses (data not shown), and was congruent with both the fossil record and current theory of vertebrate phylogeny. The great majority of nodes were supported by 100% posterior probability providing a solid statistical basis for interpretation of the tree topology. Moreover, switching between the cnidarian and molluscan outgroups did not alter the topology of the tree, and firmly established silver lamprey at the root of the vertebrate Vtgs. This finding reveals that the phosvitinless class of Vtgs, which to date has only been reported in teleosts, is more closely related to the Vtgs of vertebrates rather than insects. We propose that the phosvitinless class of Vtg is a neo-functional product of R2. The other methods of phylogenetic inference, maximum parsimony of the proteins, and maximum likelihood of the codons, also corroborated these findings.

To reconcile the multiplicity of extant Vtg genes with the "3R hypothesis" of WGD we used a discontinuous evolutionary model of the Hox gene clusters in accordance with the literature. Apparently, an ancient proto-Hox gene diverged to a para-Hox system and the Hox system, and the latter underwent several cis-duplications (intrachromosomal) yielding up to 13–14 genes before undergoing trans-duplications (interchromosomal) during R1, R2 and R3 to give rise to two, four and eight Hox clusters, respectively. Subsequent loss of genes has confused this pattern. For example until very recently zebrafish was thought to have lost the HoxDb cluster, but this has now been found in a degenerate form. The HoxDb cluster is present in the medaka and torafugu, but these species have each apparently lost the HoxCb cluster. Thus eight Hox clusters are known for the teleosts, while four are known in tetrapods, and only one Hox cluster is known in the early chordate ancestors (cephalochordates and urochordates). The presence of two Hox clusters for lamprey is congruent with scenario B proposed by Irvine *et al.*, and the observation that even though up to four clusters may exist in the Hyperoartia, they are not orthologous to the gnathostomes and hence duplicated independently. We have therefore adopted an (AB)(CD) system for effects of WGD on the Vtg gene in accordance with Hox gene nomenclature. We recognise that the results of autotetrapoloidy followed by variable rates of diploidization may not generate symmetrical trees, and that Hox gene nomenclature was established through the chronology of discovery, rather than by phylogenetic analyses as is presently proposed for the Vtg genes. We further recognise that the ancestral vertebrate containing two Hox-clusters remains a logical hypothesis.

Since the Vtg genes have the same longevity and metazoan heritage as the Hox clusters, there should be four Vtg genes in tetrapods, and eight Vtg genes present in teleosts. To date, up to four Vtg genes are known in amphibians (African clawed frog), of which two are sequenced and included in the tree. Currently three Vtg gene fragments are annotated in the genome of Western clawed frog. The multiplicity of the Vtg system in amphibians, however, is the consequence of their tetraploidy rather than the heritage of a WGD. In birds, three Vtg genes are known for the chicken, but all three are linked on chromosome 8. The major and minor Vtg genes, VtgAB2 and VtgAB3, respectively, are known to have cis-duplicated and formed pseudogenes, and are co-located ~2 Mbp upstream of VtgAB1. Interestingly, VtgAB1 does not cluster with the other avian Vtg genes, indicating that it has diverged functionally.

In teleosts, only three forms of Vtg have been found, which indicates that possibly up to five have been lost. Using the heuristic maxim of Occam's razor, we believe a more parsimonious explanation would be loss of a single Vtg gene paralogue (Vtg0) after R1. This suggestion is semi-compatible with the arguments of von Shantz *et al.*, who, based on paralogon analyses of the circadian clock *Period* genes, concluded that one paralogue (*Per4*) was lost during the early rounds of WGD. Since there is greater similarity between *Per1* and *Per2* compared to *Per3*, the loss of *Per4* may have occurred after R2 rather than R1. This remains open to interpretation, however, since von Shantz *et al.* concluded that the data did not conclusively support one pairing above others.

In the present context, the proposed loss of Vtg0 would explain the single gene in silver lamprey and the cluster of phosvitinless genes as the neo-functional product of R2. In support of this latter argument, it is noteworthy that phosvitinless Vtgs have not only lost the Pv domain, but also the ß ′ and CT domains comprising the vWFD. This is true for all currently sequenced VtgC forms, and those present in available teleost genomes.

The propensity to lose domains is not restricted to the phosvitinless class of Vtg. Current evidence for genomic Vtg genes (3-spined stickleback, medaka and zebrafish) indicates that several fragments or truncated genes are present. Among the Ostariophysi, the major Vtg genes of zebrafish, common carp, and fathead minnow (VtgAo1) have also lost the vWFD domain. In zebrafish up to seven genes, including one pseudogene have been reported, however, many more Vtg genes appear to exist in the genome. Fourteen of these genes are tightly linked on chromosome 22, while the phosvitinless gene is located on chromosome 11.

The last gene is not yet localised. We initially subjected predicted full genes (ensembl release 38, April 2006) to Bayesian analysis to find two closely related clusters, VtgAo1 and VtgAo2 and a third branch containing the more distantly related phosvitinless gene. With release 41 (October 2006) many of the earlier predicted genes are only present as fragments. We analyzed these fragments and compared them to the previously predicted genes to find that the three clusters VgAo1 (truncated), VgAo2 (complete), and VgC (phosvitinless) was supported by 100% posterior probabilities. Interestingly, the mRNA of the major gene, VtgAo1, is expressed 100–1000× higher than those of either the complete VtgAo2 or the phosvitinless form, suggesting that deleterious mutations have occurred in the cis-regulatory elements of the VtgAo2 genes. These analyses illuminate three points.

The first is that Vtgs other than the phosvitinless class can lose domains. The second is that Vtg gene forms may be present in the genome, but are not significantly expressed, while the third strengthens the probability that the phosvitinless class is indeed a product of R2 since it is located on a separate chromosome. A caveat to the third point is that the VtgC genes of 3-spined stickleback and medaka are currently located on the same group (stickleback: group VIII) or chromosome (medaka: chromosome 4) ~0.5–0.7 Mb from their VtgAa and VtgAb forms. However, since the chromosomal loci of the zebrafish VgC genes have been very recently reallocated from chromosome 15 to chromosome 11 in the latest Ensembl release (41), and the genomes of 3-spined stickleback and medaka are newly released, further analyses will be necessary to establish the definitive gene maps. Despite this caveat, maximum parsimony analysis of all genomic Vtg genes supports the topology of the tree, and the division of the major forms of Vtgs in teleosts.

Placing the R3 at the base of the crown group of teleosts as recently proposed by Crow *et al.* [31] illustrates why the timing of this WGD, or whether it in fact occurred, has remained controversial. The gene products of a WGD should show paralogous gene clustering. However, below the level of the Protacanthopterygii (salmonids), the Vtg genes do not show paralogous clustering. One interpretation for this could be the differential retention of paralogues among the different groups yielding an asymmetric tree in which the Elopomorpha, Ostariophysi and Protacanthopterygii have retained one paralogue (VtgB), and the Acanthomorpha both paralogues (VtgA and VtgB). However, with the exception of the top cluster encompassing Atlantic halibut to Japanese silago, the identities and similarities for the LvH of each of these groups (50–65%, and 70–80%, respectively) suggest otherwise. Further, since no phylogenetic method yielded orthologous relationships between the Vtgs of either the Elopomorpha, Ostariophysi, Protacanthopterygii and the Acanthomorpha or paralogous clustering of the Vtg genes immediately following R3, we suggest VtgB has been lost, and that all post-R3 teleost Vtg genes are derived from an ancestral VtgA type.

Both the Elopomorpha and Ostariophysi have duplicated their Vtg genes, but our analyses show that this has occurred in a lineage-specific manner after the WGD. Similarly, although the Vtg genes of the salmonids do show paralogous clustering, this is argued to be the consequence of an independent polyploidy event that occurred between 25–50 mya. Our current data match the first tree for salmonid Vtgs proposed by Buisine *et al.*. Interestingly, with the exception of catfish, which are also tetraploid, the eggs of salmonids are among the largest shed by any broadcast spawning teleost. The loss of a paralogue in the *Oncorhynchus* genera after separation from *Salmo* genera ~25 mya resulted in local duplications of up to 31 copies of the other paralogue, suggesting a possible compensation for the synthesis of the large amounts of yolk. Notably, the WGD at the base of the crown group of teleosts coincides with appearance of telolecithal eggs and meroblastic cleavage. With the exception of Coelacanthimorpha, all other osteichthyan fishes below the level of teleosts, have mesolecithal eggs and holoblastic cleavage. Above the level of the Protacanthopterygii, all Vtg genes show paralogous clustering giving rise to the proposed VtgAa and VtgAb paralogues among the Paracanthopterygii and Acanthopterygii. This matches the appearance of the Acanthomorpha in the fossil record 75–100 mya and their estimated origin 125 mya. Our analyses also indicate that a second putative lineage-specific duplication of the VtgA genes could have occurred with the appearance of the Acanthopterygii 55 mya. This suggests that four genes should exist in these species, but to date only dual Vtgs are known. Since we cannot preclude the existence of silent, or near silent genes as found for zebrafish (VtgAo2), this latter event may somehow represent a linked duplication or reflects a diversification of the Aa and Ab paralogues. No data are currently available for the chromosomal loci of these genes in the Acanthomorpha, and further studies will be necessary to clarify this issue. Presently we have indicated that this clustering is the result of differential cathepsin activation, since benthophils also express the VtgAa gene, but either do not, or only partially degrade the LvH during oocyte maturation.

Remarkably, the gene clustering was highly correlated to the pelagic or benthic character of the egg. Since most marine teleosts spawn pelagic eggs, and only three Vtg forms from marine pelagophils have been fully sequenced to date, we suggest that the present tree does not reflect the bias toward VtgAa type genes. The observation that only the VtgAa forms are predominantly degraded during oocyte maturation, implies that this form of the gene has evolved novel sensitivity to the developmentally regulated activation of acid hydrolases. The proposed scheme of WGD and lineage-specific duplication of the Vtg genes precisely matches the observed number of vertebrate Vtg genes in accordance with the discontinuous Hox model. For lineage-specific VtgA duplications that resulted in paralogues, we have adopted the a, b convention. To avoid confusion with the independent tetraploidization amongst the Protacanthopterygii, we used VtgAsa and VtgAsb to denote paralogous genes that were previously classified as Vtg-A and Vtg-B, respectively. The homologous Vtg genes of the Elopomorpha and Ostariophysi are preliminary classified as VtgAe and VtgAo, respectively. Further sequencing will be necessary to better establish their homology associations.

Neo-Functionalization of the Vtg Genes

The mechanism of hydration differs between benthophils and pelagophils. In benthophils, the degree of oocyte hydration is less dramatic and mostly dominated by an increase in inorganic ions and the presence of taurine, while in pelagophils, oocyte hydration is driven by an increase in the FAA content followed by inorganic ions. Recent studies of the structure and disassembly of the yolk proteins during oocyte hydration in Atlantic halibut (Finn, unpublished data) and other marine pelagophils have demonstrated that the origin of the FAA pool stems mostly from the LvH of the VtgAa paralogue. As in the more ancestral VtgABCD of lamprey and VtgAo1 of zebrafish, the VtgAb LvH may be nicked, but is essentially not degraded.

Following fertilization and formation of the yolk syncytial layer, the surviving VtgAa and VtgAb gene products (yolk proteins) are also degraded by cathepsins as substrate for the developing embryo during yolk resorption, while the FAA generated from the maturational degradation of VtgAa become the dominant catabolic substrate for energy metabolism during embryogenesis (reviewed by). In this respect the VtgAb paralogues maintain their original yolk function, but the present and previously published data demonstrate a post-duplication neo-functionalization of the VtgAa paralogues. This notion is supported by Ka/Ks ratios of 0.415 verses 0.294 for the branches that lead to the VtgAa and VtgAb clusters, respectively. Intriguingly, the VtgAa paralogue expressed in benthophils may also be partially cleaved and hydrolyzed during oocyte maturation, while the VtgAb paralogue remains intact for the developing embryo. Confirmation that the VtgAa genes have undergone neo-functionalization arises from observations that all of the yolk proteins, including those that are derived from the VtgAb clusters become exposed to the acid hydrolases during the hydration event. However, it is primarily the LvH of the VtgAa paralogues that is degraded.

Many investigations have observed the maturational disappearance of high molecular weight proteins or the appearance of the FAA pool in other species that spawn pelagic eggs, including the Elopomorpha. Indeed the FAA content of a newly fertilized teleost egg may be regarded as the signature of the pelagic egg (reviewed by). The observation that Elopomorpha also have pelagic eggs, that arise with the disappearance of a high molecular weight Lv (110 kD) and a concomitant increase in FAA indicates that the same mechanism has evolved in this group. Up to three forms of Vtg are known in Elopomorpha, but these genes are not orthologous to the Vtgs of Acanthomorpha or indeed any other teleost group. For the Japanese eel, the LvH of VtgAe1 and VtgAe2 are 98% identical, while the LvH of VtgAe3 is only 86% identical indicating that at least two forms of Vtg exist in this group. Similar to the dual Vtgs of the Acanthomorpha, only the 110 kD Lv disappears during oocyte maturation of the Japanese eel. However, due to the non-orthologous nature of eel Vtgs, we suggest that the pelagic eggs in this group have arisen by convergent evolution.

Solving the Water Problem

We argue that the solution realised by marine teleosts that spawn pelagic eggs is the generation of a large pool of organic osmolytes (FAA) that drive hydration of the oocyte while still protected within the maternal ovary. Interestingly, this adaptation is analogous to the oviparous Chondrichthyes that store the organic end-products of protein metabolism (urea and trimethylamine oxides) in their eggs. The functional significance of these mechanisms conforms to the compatible osmolyte hypothesis. Unlike charged ions, neutral amino acids, which dominate the FAA pool do not compromise enzyme function. We suggest that in teleosts, the retention of the organic end products of protein degradation was made possible by a post-R3 lineage-specific duplication of their Vtg genes, and the differential activation of acid hydrolases during oocyte maturation. Instead of converting their depolymerized yolk protein products to urea and trimethylamine oxides, they retain them in a free form as FAA, which, due to their transiently increased concentration in the yolk relative to the maternal body fluids, drive hydration of the oocyte. Once the eggs are broadcast into the sea, the formation of the virtually impermeable vitelline membrane during the cortical reaction prevents the loss of this water of life until osmoregulatory mechanisms develop during embryogenesis. The greater degradation of the VtgAa forms in pelagophils led to the highly hydrated egg and caused them to float, while the FAA comprise the major substrate that fuels embryonic development. This mechanism appears to have independently evolved in the Elopomorpha.

Oceanic Radiation and Speciation

The greater hydration associated with the spawning of pelagic eggs (>90% water) in vast numbers would have severely loaded the maternal ovary with water, and probably led to batch spawning, which is the prevalent means of reproduction in the extant Acanthomorpha. We thus argue that the pelagic nature of the egg, which arose due to neo-functionalization of the Vtg paralogues, provided the allopatric means of radiation in the oceans, while the retention of 3R gene products, latently yielded the genetic means for adjustments in pattern formation and speciation. The rapid acanthomorph colonization of the oceans occurred when competition and predation was low following the Cretaceous-Tertiary boundary extinction. Such a lack of competition and reduced predation pressure may have provided the opportunity for a flotilla of teleost invaders in the oceans.

The Pineal Gland of Fishes

Parasite Focus: Gyrodactylus

By Pete Southgate, the Fish Vet Group - Gyrodactylus is an interesting little parasite; species of which can be found infesting many different types of freshwater and marine fish as well as amphibians such as frogs. A parasitic fluke less than one millimetre in length with a direct life cycle not involving any intermediate stages or hosts. It generally infests the skin and fins of the fish, although it has also sometimes been found on the gills (it's near relative *Dactylogyrus* prefers the gills). It is also viviparous, producing live young which, rather like Russian dolls, can also produce more live young almost immediately. On microscopic examination you can sometimes see a number of generations developing within one fluke. Consequently the life cycle can be very short - between 1 and 5 days - and high numbers can therefore build up very rapidly.

Feeding mostly on the mucus and epithelial cells of the skin's surface - occasionally on blood - it attaches to the host by means of a complex hook and anchor arrangement and it is this attachment which damages tissues and causes great irritation to the fish. The fish shows obvious signs of irritation and the damaged areas can develop into significant skin ulcers and fin erosions, often with secondary bacterial and fungal infections. *Gyrodactylus* is frequently found as a mixed parasitic infestation along with parasitic protozoans such as Trichodina which also contribute to the irritation and the damage.

Having said all that, *Gyrodactylus* is not often seen as a problem in fish farms and we see it more frequently infesting aquarium and ornamental pond fish where it can cause significant disease problems.

The one species that is of concern to fisheries and fish farms is *Gyrodactylus salaris*. This is the notifiable one which is naturally widespread in the rivers of Finland, Russia and eastern Sweden draining into the Baltic Sea and which has been identified as the cause of loss of whole populations of Atlantic salmon parr and smolts in western Sweden and Norway. Epidemics of the organism occurred when it was transferred to areas where it is not naturally found and salmon exposed to the parasite for the first time had no evolved immunity (unlike those fish in the Baltic rivers which appear to be tolerant). This resulted in heavy losses of the susceptible fish and populations of salmon have been eliminated from more than twenty Norwegian rivers with the probable inevitable loss of some strain characteristics of those populations. The threat from *G salaris* to native populations of fish in Norway is considered to be so serious that all fish populations have been wiped out in river systems where the parasite has been identified. This is an extremely drastic and destructive approach and has not always proved successful in eliminating the parasite. It is, however, the only control method considered suitable and demonstrates the severity of the problem. It is absolutely imperative that *G salaris* is kept out of British waters - Scottish salmon have been shown to be susceptible to the parasite - and present European legislation is in place to control the potential introduction of the organism on live fish. It is also important that anyone possibly exposed to a risk ensures that they, and their equipment, are not contaminated with the organism.

The standard method of diagnosing *Gyrodactylus* infestation is to carry out microscopic examination of skin scrapes from the fish. The difficulty comes when trying to distinguish one species from another and the definitive diagnosis of *G salaris* infestation takes specialist expertise. The parasitologists at the Institute of Aquaculture in Stirling have done some very elegant work on distinguishing *G salaris* from other *Gyrodactylus* species and staff at the Marine Laboratory in Aberdeen have developed the expertise to characterise the organism. As a notifiable disease it is important that the regulatory authorities are notified in the case of any suspicion of the presence of *G salaris* and specialist advice should be sought on further investigation and the submission of appropriate suspicious material.

Placodermi

Placodermi is a class of armoured prehistoric fish, known from fossils, which lived from the late Silurian to the end of the Devonian Period. Their head and thorax were covered by articulated armoured plates and the rest of the body was scaled or naked, depending on the species. Placoderms were among the first jawed fish; their jaws likely evolved from the first of their gill arches. A 380-million-year-old fossil of one species represents the oldest known example of live birth. The first identifiable placoderms evolved in the late Silurian; they began a dramatic decline during the Late Devonian extinctions, and the class was entirely extinct by the end of the Devonian.

The earliest identifiable placoderm fossils are from China and date to the mid to late Silurian. They are already differentiated into antiarchs and arthrodires, along with the other, more primitive groups. Apparently placoderms already diversified into their current groups before the start of the Devonian, somewhere during the early or mid Silurian, though earlier fossils of basal Placodermi have not been discovered in these particular strata.

The Silurian fossil record of the placoderms is both literally and figuratively fragmented. All known Silurian placoderms exist today only as fragments, either scraps of armour, or isolated scales, of which some have been tentatively identified as either antiarch or arthrodire due to histological similarities. Although they have been identified, many of the Silurian arthrodire and antiarch species have not yet been formally described, or even named. Paradoxically, the best known, or rather, most commonly cited example of a Silurian placoderm, *Wangolepis* of Silurian China, is known only from a few fragments that currently defy attempts to place them in any of the recognised placoderm orders. Paleontologists and placoderm specialists suspect that the scarcity of placoderms in the Silurian fossil record is due to placoderms' living in environments unconducive to fossil preservation, rather than a genuine scarcity. This hypothesis helps to explain the placoderms' seemingly instantaneous appearance and diversity at the very beginning of the Devonian.

During the Devonian, in stark contrast to the Silurian, the placoderms went on to inhabit and dominate almost all known aquatic ecosystems, both freshwater and saltwater. But this diversity ultimately suffered many casualties during the extinction event at the Frasnian–Famennian boundary, the Late Devonian extinctions. The remaining species then died out during the Devonian/Carboniferous extinction event; not a single species survived into the Carboniferous.

Ecology and Lifestyles

Figure: *Coccosteus*

Many placoderms, particularly the Rhenanida, Petalichthyida, Phyllolepida, and Antiarchi, were bottom dwellers. In particular, the antiarchs, with their highly modified, jointed bony pectoral fins, were highly successful inhabitants of Middle-Late Devonian freshwater and shallow marine habitats, with the Late Devonian genus, *Bothriolepis*, known from over 100 valid species. The vast majority of placoderms were predators, many of which lived at or near the substrate. Many, primarily the Arthrodira, were active, pelagic predators that dwelled in the middle to upper portions of the water column. The largest known arthrodire, *Dunkleosteus telleri*, was 8–11 m (26–36 ft) long, and is presumed to have had a nearly worldwide distribution, as its remains have been found in Europe, North America and Morocco. Some paleontologists regard it as the world's first vertebrate "super-predator". Other, smaller arthrodires, such as *Fallacosteus* and *Rolfosteus* of Gogo, had streamlined, bullet-shaped head armour, strongly supporting the idea that many, if not most, arthrodires were active swimmers, rather than passive ambush-hunters whose armour practically anchored them to the sea floor. Some placoderms were herbivorous, such as the Middle to Late Devonian arthrodire *Holonema*, and some were planktivores, such as the gigantic, 7–8 m (23–26 ft) long arthrodire, *Titanichthys*.

Extraordinary evidence of internal fertilization in a placoderm was afforded by the discovery in the Gogo Formation, near Fitzroy Crossing, Kimberley, Western Australia, of a small female placoderm, about 25 cm (10 in) in length, which died in the process of giving birth to a 6 cm (2 ^{1}D $_{2}$ in) live young one and was fossilized with the umbilical cord intact. The fossil, named *Materpiscis attenboroughi* (after scientist David Attenborough), had eggs which were fertilised internally, the mother providing nourishment to the embryo and giving birth to live young. With this discovery, the placoderm became the oldest vertebrate known to have given birth to live young ("viviparous"), pushing the date of first viviparity back some 200 million years earlier than had been previously known. The arthrodire *Incisoscutum ritchei*, also from the Gogo Formation, have been found with embryos inside them indicating this group also had live bearing ability. The males reproduced by inserting a long clasper that was fused to part of the pelvic girdle, the basipterygium. Long basipterygia are also found on the phyllolepid placoderms, such as *Austrophyllolepis* and *Cowralepis*, both from the Middle Devonian of Australia, suggesting that the basiptergia were used in copulation.

It was thought that placoderms went extinct due to competition from the first bony fish, and the early sharks, given a combination of the supposed inherent superiority of bony fish, and the presumed sluggishness of placoderms. But after more accurate summaries of prehistoric organisms, it is now thought that the last placoderms died out one by one as each of their ecological communities suffered the environmental catastrophes of the Devonian/Carboniferous extinction event.

History of Study

Figure: *Homostius* and Pterichthys

The earliest studies of placoderms were published by Louis Agassiz, in his five volumes on fossil fishes, 1833–1843. In those days, the placoderms were thought to be shelled jawless fish akin to ostracoderms. Some naturalists even suggested that they were shelled invertebrates, or even turtle-like vertebrates. The work of Dr. Erik Stensiö, at the Swedish Museum of Natural History, Stockholm, from the late 1920s established the details of placoderm anatomy, and identified them as true jawed fishes related to sharks. He took fossil specimens with well-preserved skulls, and ground them away, one tenth of a millimetre at a time. Between each grinding, he made an imprint in wax. Once the specimens had been completely ground away (and so completely destroyed), he made enlarged, three dimensional models of the skulls in order to examine the anatomical details more thoroughly. Many other placoderm specialists suspected that Stensiö was trying to shoehorn placoderms into a relationship with sharks, but with more fossil specimens found, the theory of placoderms being the sister-group of chondrichthyians became accepted as fact. However, with the discovery and examination of the exquisitely preserved Gogo reef placoderm fossils, it became apparent that the placoderms shared anatomical features not only with chondrichthyians, but with other gnathostome groups, as well. For example, Gogo placoderms show separate bone for the nasal capsules which are incorporated into the braincase of both sharks and bony fish. Because of these new insights provided by the Gogo Reef specimens, coupled with the fact that placoderms also share anatomical features only with the jawless osteostracans, the theory that placoderms are the sister group of chondrichthyians has been replaced in favour of the theory that placoderms are a group of stem gnathostomes, in other words, they are the sister group of all other known gnathostomes.

Ostracoderm

Ostracoderms ("shell-skinned") are any of several groups of extinct, primitive, jawless fishes that were covered in an armour of bony plates. They belong to the taxon Ostracodermi, and their fossils are found in the Ordovician and Devonian Period strata of North America and Europe. They were often less than 30 cm (1 ft) long and were probably slow, bottom-dwelling animals. Another innovation of ostracoderms was the use of gills not for feeding, but exclusively for respiration. Earlier chordates with gills used them for both respiration and feeding. Ostracoderms had separate pharyngeal gill pouches along the side of the head, which were permanently open with no protective operculum. Unlike invertebrates that use ciliated motion to move food, ostracoderms used their muscular pharynx to create a suction that pulled small and slow moving prey into their mouths.

Ostracoderms were the first fossil fish to be discovered. The Swiss anatomist Louis Agassiz received some fossils of bony armoured fish from Scotland in the 1830s. He had a hard time classifying them as they did not resemble any living creature. He compared them at first with extant armoured fish such as catfish and sturgeons but later realising that they had no movable jaws, classified them in 1844 into a new group "ostracoderms" which means "shell-skinned". Ostracoderms existed in two major groups, the more primitive heterostracans and the cephalaspids. The cephalaspids were more advanced than the heterostracans in that they had lateral stabilisers for more control of their swimming. After the appearance of jawed fish (placoderms, acanthodians, sharks, etc.) about 420 million years ago, most ostracoderm species underwent a decline, and the last ostracoderms became extinct at the end of the Devonian period. More recent research indicates, however, that fish with jaws had far less to do with the extinction of the ostracoderms than previously assumed.

The Subclass Ostracodermi has been placed in the Superclass Agnatha along with the extant Subclass Cyclostomata, which includes lampreys and hagfishes. The term does not often appear in classifications today because it is paraphyletic or polyphyletic. However, "ostracoderm" is still used as an informal term for the armoured jawless fishes of the Paleozoic.

Major Groups

Some major groups of ostracoderms are as follows.

- Pteraspida
- Thelodonti
- Anaspida
- Galeaspida
- Pituriaspida
- Osteostraci

Cyclostomata

Cyclostomata is a group of chordates that comprises the living jawless fishes: the lampreys and hagfishes. Both groups have round mouths that lack jaws but have retractable horny teeth. The name Cyclostomata means "round mouths". Their mouths cannot close due to the lack of a jaw, so they have to constantly cycle water through the mouth.

Possible Relationships

This taxon is often included in the paraphyletic Class Agnatha, which also includes several groups of extinct armoured fishes called ostracoderms. Most fossil agnathans, such as galeaspids, thelodonts, and osteostracans, are more closely related to vertebrates with jaws (called gnathostomes) than to cyclostomes. Cyclostomes seem to have split off before the evolution of dentine and bone, which are present in many fossil agnathans, including conodonts.

Biologists disagree about whether cyclostomes are a clade. The "vertebrate hypothesis" holds that lampreys are more closely related to gnathostomes than they are to the hagfish. The "cyclostome hypothesis", on the other hand, holds that lampreys and hagfishes are more closely related, making cyclostomata monophyletic. Most studies based on anatomy have supported the vertebrate hypothesis, while most molecular phylogenies have supported the cyclostome hypothesis.

There are exceptions in both cases, however. Similarities in the cartilage and muscles of the tongue apparatus also provide evidence of sister-group relationship between lampreys and hagfishes. And at least one molecular phylogeny has supported the vertebrate hypothesis. The embryonic development of hagfishes was once held to be drastically different from that of lampreys and gnathostomes, but recent evidence suggests that it is more similar than previously thought, which may remove an obstacle to the cyclostome hypothesis. More data may be needed to be more confident which theory is correct.Pigmentation in marine fishes

The fantastic array of colours found in tropical reef fishes is what draws many of us to become involved in marine aquarium keeping in the first place. Vibrant images of grand structures like the Great Barrier Reef, with its collection of sessile and motile inhabitants are what typically pique our fascination with marine creatures. But consider for a moment that the spread of colours exhibited amongst these animals is not that different from an artist's palette. The blending of a limited number of primary colours yields new paint mixtures for the artist. Analogously, in the great oceanic food chain there is a vast source of basic pigments formed by the primary producers (photosynthetic bacteria and algae) which through feeding, metabolic processes and conversions eventually end up within the tissues of fishes, giving them their colourations. But unlike an artist's palette, the basic pigments of the lowly primary producers have a greater purpose than for mere visual stimulation or esthetic appeal. They are vital to the nutritional content of marine food sources and to the very maintenance of fish health. In essence, they are part of a complex of nutrients found within natural oceanic foods and therefore should be a consideration for any aquarist who is dedicated to providing the best of care for their captive marine fishes.

The natural pigments, distributed in the fatty-tissues of marine fishes and invertebrates, mostly begin as products synthesized within the tissues of photosynthetic bacteria, algae, and higher plants. These yellow-to-orange-red pigments, called *carotenoids*, of which there are over 600 hundred different chemical formulations, are typically divided into two subgroups, the *carotenes*, or hydrocarbon carotenoids, and the *xanthophylls*, or oxygenated carotenoids. The carotenoid family of pigments, while not responsible for all fish colourations, is the most widespread of pigments in marine ecosystems. Adding to the great variety of carotenoid pigments already in existence, many organisms have the physiological ability to chemically alter pigments obtained from their food sources, thus producing new colour variations. Also, carotenoids deposited within the fatty-tissues of animal structures interact by overlaying base colours. An example is the Pink Flamingo, a bird that obtains its pink colouration from minute crustaceans it sifts from shallow waters with its down-turned bill. Captive flamingos fed a diet lacking in carotenoids lose their proper colouration and turn white. Another more familiar example to marine aquarists, is the Yellow Tang (*Zebrasoma flavesens*) which often fades and develops white patches when deprived of algae or other greenery in its diet. For this reason many prepared foods specifically designed for marine fishes contain sources of xanthophylls like algae.

There are a series of physical mechanisms within the cellular and anatomical structures of fishes that enable them to reflect and to refract light and more importantly, to display colours obtained from dietary sources. The physical characteristics of fish skin are part of the key to how this is all accomplished. First, there is the epidermis. The epidermis is most often transparent and is the very thin outer layer of the skin. The epidermis can contain mucous glands, sensory cells (taste buds and alarm cells) and pigment cells. Scales, if they are present, lie below the epidermis in pockets as part of the dermis. (Fish scales are generated in the dermal layer and are not shed like those of reptiles, which are epidermal in nature). The dermis contains pigment cells of various kinds. Pigments cells in general are called chromatophores. Individual chromatophores may contain several types of pigments. Each pigment is produced or contained in separate cellular organelles called chromatosomes. The transparency of the scales and the thinness of the epidermal layer are what allow the dermal pigments of the chromatophores to show through the skin for display.

Basic melanin pigments like black and brown, interestingly enough, are not obtained from dietary sources like bacteria, algae, or plants. Instead, they are protein-based pigments manufactured within pigments cells called melanophores. The almost all-black colouration of a juvenile Domino Damsel (*Dascyllus trimaculatus*) is one example of melanin pigmentation. The pattern of vertical bars in the Convict Surgeon (*Acanthurus triostagus*) and the complex arrangement of the Clown Triggerfish (*Balistoides conspicillum*) are others. But whether the pigment is manufactured in the cells or concentrated from a dietary source, it is the physiological mechanism of chromatophores that allow for the changing of colour, pattern, and intensity in fishes. The various intracellular pigment-containing organelles can be concentrated in the cell or diffused throughout it. If they are diffused, the colour is presented as more intense, and if they are concentrated the colours is muted. This cellular mechanism enables fishes to alter the intensity of colour according to background, lighting conditions, and behavioral mood.

Another source of perceived colour in fishes is refraction. Refracted or scattered colours are not actually the result of pigments but are created by crystalline intracellular deposits. Concentrated within cells called iridiophores, these chemicals bounce light in such a way that the reflected colours change depending on the angle of reflection.

Figure: The false percula clown (Amphiprion Ocellaris) is a darling of the captive breeding industry. Early breeders learned quickly that if breeding clownfish and their fry were denied the proper pigments in their diet, not only would they lack bold colouration, but the fry would fail to develop and mature properly. Thankfully for commercial interests, pigment concentrates are available as food additives.

The deposition of carotenoids is not limited to the skin of fishes. Nor is the basic colouration provided by carotenoids the only biological function of the pigments. Carotenoids are fat-soluble and can be deposited in tissues throughout a fish's body, including skin, muscle, fins, eggs, gonads, milt, and sensory organs like the eye. The layering of pigments in the skin, along with the cellular manipulation of pigments, allows fish to change colour for cryptic, behavioral, or communication purposes. Many fish species in spawning condition exhibit colour changes which can be attributed to nutritional triggers such as a surplus of a particular food source. Carotenoids within females are passed on to eggs and fry and the bright red or pink colours typically found in fish eggs can be a signal for fertilization or for parental care to begin. A major function of carotenoids, other than colouration, is as a provitamin, or precursor to the synthesis of vitamins. An important nutrient for most animals is vitamin A. Vitamin A cannot be synthesized by plants, but it is essential for the proper vision, growth, immunity, and reproduction of fishes. Fishes do have the ability to chemically convert some carotenes into vitamin A, often by splitting of a molecule. Another example of the nutritional value of pigments is astaxanthin. Astaxanthin is a red pigment from the xanthophyll oxygen-carrying group of carotenoids and is considered to be a powerful antioxidant... one hundred times more powerful than vitamin E. Research suggests that astaxanthin functions as a intracellular oxygen reserve and fishes and invertebrates fed on a diet containing optimal levels of astaxanthin show greater survivability rates when subjected to poor oxygen environments. Lastly, as a pigment, astaxanthin provides a degree of photo protection for reef fishes and invertebrates, which both must survive in clear waters flooded with ultra violet light. Proper deposition of such pigments in the eyes and skin of marine creatures is necessary for long-term survival and maturation.

Fulfilling the nutritional requirements of marine aquarium fishes for carotenoid pigments (just like their overall nutritional requirements) in their prepared foods is inherently connected to the natural feeding methods of the fish species in question. For example, there are the grazing species like damselfishes, tangs, and blennies, which browse continuously on algae. They obtain the essential carotenoid (xanthophylls) pigments from consuming green, red, and blue-green algae. Certain species of damselfishes (*Stegastes* and Garabaldi), keenly "farm" small patches of filamentous algae in the natural habitats, fiercely keeping away any intruders. Many species of Tang (*Acanthurus* and *Zebrasoma*) school over reefs picking at targeted filamentous algae. Blennies like the Jewelled Rockskipper (*Salarius fasciatus*) rasp surrounding surfaces of micro-algae growth. To achieve a proper regiment of feeding these types of fishes, different sources of food must be employed. Relying on surface algae growth in an established marine aquarium to support grazers may not be adequate. Grazers are quick to reduce excess algae growth.

Remember that herbivorous material takes longer to digest, so grazers must feed almost constantly in order to remain healthy. Unfortunately, this makes herbivores prone to malnutrition if food is in short supply. But a diet of algae growth can be supplemented with sheets of dried macro-algae specifically prepared for marine aquarium fishes, or even with sheets of Nori, which is used in wrapping sushi. Feeding spirulina (blue-green algae)- based flake foods is also an excellent addition to a grazer's diet. If brine shrimp are offered to the grazing fish and if they are accepted eagerly, then the aquarist can easily rear the brine shrimp on algae powder so the needed carotenoids will be passed on to the fish when consumed. (Gut-loading, as it is called, is a common practice in other hobbies like reptile-keeping) but natural and more direct sources of algae are always better, since fish can graze on them at will throughout the day instead of at the designated feeding times. Aquarists who utilise an algae-scrubber filter can remove the scrubber plates, the surfaces covered with the micro-algae growth, and place them directly in the main aquarium where the fish can nibble them clean. Or if a lighted sump or refugium is utilised, different types of filamentous algae can be grown and harvested for the fish.

Figure: The ever-popular yellow tang (*Zebrasoma flaviscens*) will become pale and develop patchy colouration if it does not receive a diet rich in natural pigments.

For many species of small fishes, which feed on drifting zooplankton in their natural habitat, and rely on this as their primary nutritional source, the proper prepared foods in the confines of a marine aquarium is paramount. Fishes like *Anthias*, Clownfishes, Dottybacks, Dartfishes, and *Grammas* pick at a variety of drifting particular foods like crustaceans, invertebrate and fish larvae, and planktonic eggs, to name a few. The pigments obtained from these foods begin in the primary producers and are passed along in the food chain to the zooplankton and then on to the zooplankton feeders. Fishes that employ this feeding method are accustom to some level of water movement and require several small feedings per day. Small crustaceans like mysid shrimp and brine shrimp, or colour enhancing flake foods that contain carotenes are ideal for these fish species.

Finally, there are the marine fish species that come under the general grouping of piscivores. (The distinction here is that the carnivore may eat animal flesh regardless of size or type, from a small crustacean to a coral polyp to another fish, whereas the piscivore is defined as specifically eating fish). This behavior, which can clearly be a disadvantage in a tank of ornamental fishes, is found among species of Lionfish, groupers, Moray Eels, Frogfishes, and of course, sharks. To accommodate a piscivore's dietary needs, meaty pieces of marine fish and crustacean should be offered a few times per week. Any live foods offered, regardless of whether they are fish or crustaceans, should be well reared on carotenoid rich foods. But taking into consideration the fact that obligate carnivorous fish eat less frequently than obligate herbivorous fish and are often prone to malnutrition, vitamins enrichment of the food offered as well as loading of the food with algae may go a long way to enhance a piscivore's diet and health. (Even strict carnivores like lions and tigers obtain plant material in their diets from devouring herbivorous prey. The gut contents, or undigested plant material, of the deceased herbivorous animals are passed on to the carnivorous predators). All the nutritional requirements of piscivorous fishes including important carotenoids are passed along in the food chain from their algae eating and plankton eating prey. So, adding some type of carotenoid enhancement to the foods offered to piscivorous fishes is helpful in fulfilling their complex nutritional requirements and in maintaining their pleasing colours. *[Editor's note: Prey items for larger predators can literally be stuffed or injected with commercial fish foods, vitamins or other supplements.]*

In the end, pigmentation in tropical marine fishes is a subject more involved than the simple display of colours and patterns. The vast array of colours exhibited amongst reef fishes has a far greater purpose than for visual stimulation, cryptic colouration, or warning displays. There are intricate microbiological and physiological functions associated with the pigments deposited in the tissues of fishes. The various pigments are essential for proper growth, immunity, and reproduction. And as marine aquarists, we must appreciate the fact that a healthy level of these pigments must be maintained in the diets of our aquarium fishes. The natural food chain provides for the distribution of pigments in the wild, all the way from the bottom primary producers to the top consumers. But within the confines of our home aquariums, our diligence and intervention is required in providing these vital nutrients.

What are Endocrine Disruptors?

Endocrine disrupting chemicals are a category of environmental contaminants which have received wide-spread attention due to the risks they pose to freshwater fish systems. The easily accessible pathways which allow them access and their lasting persistence within freshwater systems only stand to enlarge this threat. The increasing presence of intersexuality in contaminated fish populations is one of such effects caused by ED contamination which has serious repercussions for species' reproductive success and therefore species success as a whole. This study aims to explore the range of influence that ED's have on fish populations, and explores current legislative initiatives to remediate their effects, whilst proposing methods of it's own.

Biological Repercussions of Endocrine Disrupting Chemicals: The greatest effect that endocrine disrupting chemicals have on freshwater fish species is the disruption of their biological reproductive success. As a result of the this disturbance, patterns of extinction and endangerment may emerge. The factors which lead to reproductive complications are all dependant on the amount of exposure, the length of exposure, and the timing of exposure to ED's relative to development. Intersexuality, in which a fish exhibits the presence of both male testes and female ovary structures, is one of the chief symptoms of ED contamination. Testosterone is the masculine hormone in fish species, while estrogen (a main endocrine disruptor) is the typical feminine hormone. The level of testosterone hormones are chiefly important in male sexual development during the early stages of fish life cycles. The presence of abnormal levels of estrogen in the water as a result of ED contamination, can lead to disturbed sexual differentiation in fish through their increased influence over testosterone during the fish's early stages of maturation. An effect of this disturbed masculinity in fish is the malformation of ovarian and sperm ducts which leads to reduced sperm quality, thus reducing the species fertilization capabilities. ED's also effect female fish's fecundity, or amount of eggs produced.

Another widely studied effect that endocrine disruptors have on fish species is an increased amount of vitellogenin in male fish blood plasma. Vitellogenin is a female-specific protein which factors into egg-yolk production and controls the growth of unfertilized reproductive cells (oocytes). The production of vitellogenin is monitored by estradiol levels within the liver. (Rotchell) As a result of this link, the increased presence of ED's, especially estrogenal hormones, causes male fish to exhibit this female specific gene, causing disturbed reproductive success.

Another biological process which is effected by ED's is zonagenesis. Zonagenesis is another important reproductive process which is controlled by estradiol levels. (Rotchell) Through zonagenesis, certain proteins are synthesized which come to form the eggshell, or membrane, which in turn supports the embryo. The influence of ED contamination disrupts this process, leading to higher birth failure, and fecundity rates in fish species.

Fish patterns of sexual selection are also affected by ED influence. Studies have found that the characteristics that males display which are most successful in mating selection can be amplified in inter-sex males and males with reduced expressive sperm counts. This factor gives the disturbing influence of ED's even more potential for impacting the success or failure rate of fish species.

How ED's Work

Endocrine disrupting chemicals are introduced into freshwater systems through the use of hormonal pharmaceuticals, primarily the female birth control pill, waste-water treatment plant effluents, industrial inputs, agricultural runoff of herbicides, pesticides, and fungicides, septic system inputs, and other non-point source pollution. Effluents from waste-water treatment plants have been acknowledged as a main route for the introduction of ED's into freshwater systems, especially in Minnesota. Cases of endocrine disruption in fish species have been reported in the effluent channels of the Minneapolis-St. Paul Metropolitan Waste-Water Treatment Plant, affecting local carp and walleye population within those systems. (Lee)

The harmful effects of endocrine disrupting chemicals are wrought through interference with the natural routine of endocrine system interchanges. ED's can disturb this cycle by binding or blocking different estrogen receptors, changing natural hormone levels, dislodging a hormone from a carrier protein which in turn makes the protein useless for transfer, disrupting hormone synthesis, and through affecting other hormone receptors of the cell membrane. In regards to intersexuality in fish, particular consideration should be made with reference to testosterone and estrogen receptors. Androgen receptors control testosterone levels in fish, and are very important in male sexual development. In the endocrine system, estrogens and androgens each bind with their respective receptors, in turn then binding to specific DNA recognition sites which control the target genes to be transcripted. When fish species are exposed to ED's this process can be inhibited, or the pathway altered to allow non-typical proteins to be found in male specimens. Vitellogenin is a great example of this, in which a female-specific protein's synthesis is encouraged in males by increased estradiol levels. Some ED's, such as PCB's can even be metabolized to mimic estrogen molecules, and allowed to bind with estrogen receptors. (Rotchell) The success of ED bondage within the endocrine system is subject to natural variation, and species differences.

Descriptions of Three Different ED's

A. Steroidal Estrogens: In freshwater systems there are three prominent types of steroidal estrogens present: estradiol (E2), estrone (E1), and ethinylestradiol (EE2). Estradiol and estrone are natural estrogens, and are found higher in concentration in freshwater systems that ethinylestradiol. However, it has been found that the potency of EE2 makes it's exposure more of a problem, and most studies have been focused on it's effects. EE2 is prominently used in oral contraceptive birth control pills, and finds it's route into freshwater systems through wastewater containing the wastes of humans who consume it. EE2 has a longer half-life than natural estrogens and is easily biocencentrated in fish. It has been found that it's chemical potency in fish is ten to fifty percent higher when exposed during development than that of the natural estrogens. The negative biological consequences this ED has for fish also can occur from relatively low levels of exposure. At only .1ng/l, EE2 can encourage increased vitellogenin production in fish, and at .1-.15 ng/L can disrupt sexual development and differentiation. (Nash) EE2 also has a potent effect on fertilization; in a study conducted on Zebrafish it was found that exposure to EE2 reduced fertilization success to 51.8%, down from the control groups success rate of 95%.

B. Atrazine: Atrazine is another prominent ED found in freshwater systems. A popular herbicide, atrazine is the most common pesticide contaminant of surface water. It's biological repurcussions have been well noted, and some groups, like the European Union, have even banned it's use. Atrazine is a very enduring contaminant, owing to a half-life of thirteen to two-hundred and sixty days. It also has a high mobility rate, finding it's way into freshwater systems through surface run-off. All these factors combine to make Atrazine a considerate threat. This ED, like many others, increases estrogen production in exposed species causing feminization and heightened possibilities of reproductive dysfunction. Atrazine has also been found to increase aromatase levels in fish species. Aromatase is and enzyme linked to estrogen production, and as such a significant factor in reproductive development. Increased levels of aromatase in male fish lead to increasing numbers of androgens being aromatized, a process which turns testosterone into the female estrogen hormone. (WuQiang)

C. Polychlorinated Biphenyls: The third ED contaminant group we will focus on are polychlorinated biphenyls, or PCB's. PCB's are a very popular and widespread environmental contaminant, and has seen a lot of attention due to the risks it poses for human populations. PCB's were popularly used in many coolants, plasticizers, adhesives, and other products, and transfer through the atmosphere. They do not degrade very easily, and are found in high levels in urban pollution and soil concentrations. PCB's also have a high solubility in fats, making them easier to bioconcentrate. PCB's have been known to have carcinogenic effects in both humans and animals, and much legislation has been passed due to these health concerns. They can also have an effect on estrogen levels, especially when metabolized. PCB's can be metabolized to resemble estrogen molecules and can then bind to certain gene receptors which disrupt sexual development through the increased presence of estrogen in the endocrine system.

Materials and Methods

Sexually mature goldfish were collected from commercial dealers (mean weight 45±2.5 gm) and were kept in an aquarium until the experiment began in April. Aerated and dechlorinated water with flow rate of 1.5 L min^{-1} 9 ppm dissolved oxygen, 7.8 pH and 102 mg as $CaCO_3$ total water hardness was used. The experiment lasted for 40 days. The fishes were fed twice daily with a commercial fish meal based extruded diet (diametre/mm; 55% Crudeprotein 14% Crude lipid, 4296 Cal g^{-1} diet gross energy) at 3.5% body weight at 10:00 am and 14:00 pm. One hour after feeding uneaten feed was removed by the sandpipe at the bottom of the tank. Fishes were transferred in a glass aquarium size 4 x 4 x 2 feet. Two aquariums were used, one was filled with Yamuna River water and twenty fishes were kept in each aquarium. In second aquarium fishes were kept in ground water (pure and unpolluted) in natural conditions. Bottom of the aquarium was covered with sand, submerged plants and stones. Blood samples were collected by caudal peduncle excision.

R.B.C. and W.B.C.count through the blood from the caudal vein from control and treated fishes was collected for haematological investigation. R.B.C. and W.B.C.by (Wintro, 1967) mean corpuscular haemoglobin concentration (MCHC) were determined according to Dacie and Lewis (1975), haemoglobin content (Vankampen, 1961).

Figure: Effect of polluted water of River Yamuna on various parameters of Goldfish polluted water (Group I) and unpolluted water (Group II).

Parameters	*Group-I*	*Group-II*
W.B.C. Count (mm^3 x $10^{3)}$	3.42 ± 0.36	5.88 ± 0.24
R.B.C. Count (10^6 $mm^{-3)}$	1.02 ± 0.09	1.86 ± 0.1
Haemoglobin(g/100ml)	4.1 ± 0.59	6.98 ± 0.38
MCHC %	8.6 ± 0.26	12.2 ± 0.15

All data are presented Mean ± Sem for GSI %, R.B.C. Count, W.B.C. Count, Hb, and MCHC %. All data are differed significantly P<0.05.

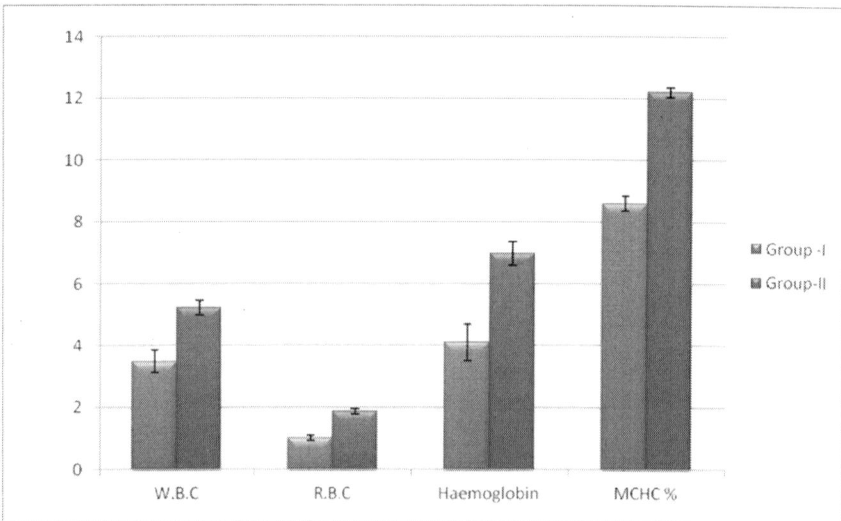

Figure: Showing values of different blood parameters of Goldfish, error bars indicates Standard error of mean.

Fishes are in indirect contact with their surrounding environment and any change in the environment will be reflected as changes in their physiological processes and survival. Fishes possess shorter development time compared to mammalian species (Peakall, 1992).

Fish are one of the most widely distributed organisms in an aquatic environment and being susceptible to environmental contamination may reflect the extent of the biological effects of environmental pollution in waters. Monitoring of blood parameters, both cellular and noncellular may have considerable diagnostic value in assessing early warning signs of pesticide poisoning (Pant et al., 1987). It is well known that water pollution is an issue beginning to receive more attention recently. Discharging into the rivers, discharging of sewage of fish growing fields are major sources of pollution. But no considerable steps have been taken for fish growing fields in the country and the sewage water pouring into rivers. Since rivers constitute main sources of fresh water used in agriculture, drinking, and industrial applications, everything that causes the water pollution can be considered as an important issue to be studied (Smith,1997). In this study when fishes were held in polluted water in aquariums (Group-I and in ground water (Group-II), the changes in the morphological appearances and the change in haematological parameters of the fish were observed.

In group I, 5 fish died due to pollutants present in river water so the survival % in group I was less than group II. R.B.C., W.B.C. count, MCHC % and Haemoglobin content also decreased significantly ($p<0.05$) in comparison with group II in which fishes were held in unpolluted ground water. Our results concluded that water pollution in river and other water bodies is very harmful for fishes because of pollutants present in water, polluted water reduced survival, blood parameter levels in Goldfish. Water pollution is an issue beginning to attract more attention recently; discharging of sewages, effluents from houses, industrial water, industrial wastes. Pesticides from fields also polluted rivers and lakes and this polluted water is harmful for fishes to survive and reproduce so it is our duty to stop polluting the water to save fishes. Water pollution is a major global problem which requires ongoing evaluation and revision of water resource policy at all levels. It has been suggested that it is the leading worldwide cause of deaths and diseases.

River Sewage Pollution Found to be Disrupting Fish Hormones

New research has strengthened the case that polluted river water is harming fish's ability to reproduce by disrupting their hormonal, or endocrine, systems.

Scientists found that sewage treatment works are releasing significant volumes of chemicals that inhibit the function of testosterone, the male hormone, into British rivers. They call these chemicals 'anti-androgens'; such pollutants might include compounds used in pharmaceuticals, cleaning products, cancer treatments and pesticides.

These anti-androgens are likely to contribute to fish feminisation - a process whereby male fish develop some female reproductive characteristics, reducing the quality and number of sperm they produce and probably also their ability to compete to pass on their genes. In other studies exposure to anti-androgens has also been linked to damage to human reproductive health. Scientists already suspected that the female hormone estrogen could cause feminisation in fish, but it turns out that a whole cocktail of chemicals acting in combination is likely to be behind the problem.

'Our research shows that a much wider range of chemicals than we previously thought is leading to hormone disruption in fish,' says Professor Charles Tyler at the University of Exeter, one of the paper's authors. 'This means that the pollutants causing these problems are likely to be coming from a wide variety of sources.'

Tyler adds that 'the circle is getting tighter' in establishing the hypothesis that similar pollution may also be causing male fertility problems in humans, though he notes that no such connection has yet been proved. He explains that in the lab exposures to estrogenic chemicals at the concentrations found in the natural environment do not seem to cause problems of the severity seen in wild fish; when the oestrogens are combined with anti-androgens the mixture is likely to produce the effects seen in polluted rivers.

The study was carried out by the Universities of Brunel, Exeter and Reading with the Centre for Ecology & Hydrology, with statistical modelling support from Beyond the Basics Ltd, and appears in *Environmental Health Perspectives*. It is based on more than 1000 fish sampled over three years from 30 rivers across England, and on samples of effluent from 43 sewage works.

'We have been working intensively in this field for over ten years,' says Dr Susan Jobling at Brunel's Institute for the Environment, lead author on the paper. 'The new research findings illustrate the complexities in unravelling chemical causation of adverse health effects in wildlife populations and re-open the possibility of a human-wildlife connection in which effects seen in wild fish and in humans are caused by similar combinations of chemicals.'

Something in the Water

The University of Exeter is now working with chemists at the University of Sussex to establish exactly what the chemicals in question are, and hence to understand how they are being released into waste water. This will involve both chemical testing of effluent from sewage treatment works and experiments with fish at the works' outlets. Fish accumulate these chemicals in their bile, so can be used to make identifying them easier.

The scientists will also be working with regulators and the water industry to test the chemicals' effects of these anti-androgens in combination with estrogen and related female hormones to get a realistic analysis on their combined effects in our river systems.

Earlier research from Brunel and Exeter scientists showed how female hormones like estrogen are leading to the feminisation of male fish. Found in the contraceptive pill as well as some industrial chemicals, estrogen and associated chemicals enter rivers via sewage treatment works, causing reproductive problems in fish by reducing fertility and in some cases even causing male fish to partially change sex. One common cause of fertility problems in male humans is called testicular dysgenesis syndrome. This syndrome is known to be caused by exposure to anti-androgens in lab mammals.

The fact the team has proved that anti-androgens as well as estrogen are being emitted in significant quantities into British rivers certainly provides indirect support for the hypothesis that pollution is behind at least some of the increase in human fertility problems. The researchers say more work is now needed to test this hypothesis.

'In vertebrates the hormonal system is highly conserved; ours is comparatively close to that of fish,' Tyler notes, though he also points out that the kinds of exposure to river pollution a human could face is very different from what a fish experiences swimming in a soup of 100,000 chemicals contained within effluent discharges - in some cases throughout the fish's life.

Other researchers also call for caution in connecting the cases of fish and humans. 'Are anti-androgens part of the mix of chemicals that could be disrupting wildlife and humans? Yes,' says Dr Andrew Johnson, an environmental scientist at the Centre for Ecology & Hydrology in Wallingford who has spent a decade working on endocrine disruption.

'This paper shows that what we already knew was happening to humans is also happening in freshwater,' Johnson adds. He explains that research on the hormone-disruption affecting humans has tended to focus on a group of chemicals called phthalates, to which humans are believed to be exposed primarily through cosmetics, and argues that it's not clear that chemicals found in sewage effluent are likely to return to the water supply to affect humans.

Whether the chemicals can do so will depend on exactly which anti-androgens are discovered to be present. If those present are found to be particularly hydrophilic, or water-soluble, it's possible they could get through filtration procedures and return to the drinking water supply, but these filters efficiently remove most pollutants.

Inbreeding and Brood Stock Management

Inbreeding is a genetic concept that most aquaculturists have heard about, but few understand what it is or know how it can be used or abused. Inbreeding is one of the few natural biological processes that evokes deep-seated prejudices and invokes legal sanctions; as a result, it is probably the least understood and least appreciated aspect of animal and plant breeding. Most have heard of inbreeding because of legal and moral laws against certain consanguineous ("blood-related") marriages and because extension agents and others have repeatedly preached about the evils of inbreeding. However, inbreeding is no different from any other piece of technology. If used properly, it can improve things; if used improperly, it will make them worse.

Inbreeding has been a concept that has plagued and intrigued humanity for thousands of years. The Torah, Bible, and Koran all contain proscriptions against incest. Most societies have laws pre-dating constitutional rights that prohibit marriages between certain relatives, although feudal rulers routinely ignored these laws, because it was a prudent way of consolidating money, land, and power.

The major reason why inbreeding was considered to be immoral was that no one knew anything about genetics, and consequently no one knew how inbreeding worked. It was easy to observe that when close relatives mated, some offspring were deformed or were weak and died. Since genetics was unknown, people associated the deformities, etc. with the act of incest itself and thus concluded it was evil and the deformed offspring were some form of divine punishment.

Our aversion to consanguineous marriages and abhorrence of incest is probably rooted in our evolutionary past and is controlled by our genes. Studies with several species of animals have shown that various behaviours, such as dispersal patterns-either forced or voluntary-of juveniles who are about to become sexually mature prevent parent-offspring, brother-sister, or first cousin matings which produce high levels of inbreeding. Since inbreeding produces some subviable or inviable offspring, genetically controlled behaviour which prevents such matings will be favoured by natural selection.

Prohibitions and prejudices against inbreeding were so deeply rooted in Europe that as recently as 200 years ago European animal breeders were told that they should not use inbreeding in their brood stock management plans because it was against the Laws of God and Nature. However, animal breeders ignored these injunctions when they discovered that inbreeding was a breeding technique that could be used to develop new breeds, improve herds, and produce outstanding animals. Once the science of genetics was established, geneticists were finally able to explain why some inbred offspring were deformed and were able to show that this was a problem of heredity, not of morality. They were also able to explain how inbreeding produced better animals and plants. An understanding of the genetics of inbreeding enabled geneticists to devise regular systems of inbreeding so animal and plant breeders could use inbreeding to improve growth rate and other phenotypes in a reliable and predictable manner.

Inbreeding is the mating of relatives, or the mating of fish more closely related than the population average (which is another way of defining relatives). Some geneticists want to restrict the definition of inbreeding and say it is the mating of "close relatives," and that means the mating of two individuals more closely related than second cousins. The restricted definition might be more practical as a breeding definition, because it is based on the amount of inbreeding in the offspring that are produced by a mating. However, the mating of any relatives, whether they are closely related or distantly related, is inbreeding, and it will produce inbred offspring.

Inbreeding is, for all practical purposes, the opposite of crossbreeding. While inbreeding is the mating of individuals more closely related than the population average (between relatives), crossbreeding is the mating of individuals less closely related than the population average (between individuals from two populations).

Inbreeding, along with selection and crossbreeding (hybridization), is one of the three major breeding programmes that have been traditionally used to improve livestock and plants. While selection and crossbreeding are more widely known and appreciated, inbreeding is a powerful breeding technique that has been used to: establish new breeds and varieties; improve the results of selection; create better brood stock; and improve the results of crossbreeding programmes.

There is a wealth of information about inbreeding in livestock and laboratory animals and an even greater treasure trove for plants. However, comparatively little information exists for farmed fish; most of the information that does exist is spotty or consists of warnings about why inbreeding should be avoided and explanations about how it can be prevented.

While inbreeding is an important breeding programme that can be used to improve a population when it is planned and directed, unplanned and uncontrolled inbreeding can ruin a population through something called "inbreeding depression," which is a decrease in growth and viability coupled with an increase in abnormalities. Genetic aspects of fish stock management are still in their infancy. Many farmers are still raising stocks that are either wild or only a few generations removed from the wild. Most farmers worry only about the environmental aspects of management-nutrition, water quality management, disease control and prevention-and ignore the genetic aspects. Since the modern science of aquaculture is still in its infancy, these management decisions make sense, because the best way to raise a crop has often not been quantified. Vast and immediate improvements can be obtained by refinements in fertilization regimens, feeding practices, or water quality management. These changes are often very cost effective because the changes are inexpensive.

The genetic aspects of fish farming have received far less attention because they require sophisticated managers, are long-term forms of management, require extra facilities, and add additional financial burdens; furthermore, the results are often unpredictable. While great improvements are possible, they are usually more costly than environmental improvements, and the return is not immediate.

Those who are interested in incorporating genetic improvements into yearly work plans usually want to conduct selective breeding or crossbreeding programmes; inbreeding is ignored. There are two probable reasons for this: The first is the deep-rooted prejudice against inbreeding. The second is the warnings against inbreeding that come from fish geneticists and aquaculture extension agents.

These warnings have been sounded because fish farming is a form of animal husbandry where inbreeding can be a more serious problem than it is with livestock. For the most part, fish farmers work with small populations. When new stock is acquired, the number of fish acquired is usually small because: shipping fish is a lot of trouble; it is costly, especially since water is heavy; and it is difficult to ship fish great distances. Consequently, many shipments contain only a handful of fish. Often, the fish that are acquired come from one or two matings, which means the genetic size of the population is quite small. Finally, fish are highly fecund compared to traditional livestock, which are capable of producing only one to a dozen offspring each year. Because of this, there is often a great temptation to spawn as few fish as possible in order to save money and labour and to lower the cost per fingerling produced. The fecundity of some fish is so great and the size of many farms is so small that a single spawn will often produce enough fish to stock many farms.

A number of warnings against inbreeding have been published because aquaculture is a unique form of animal husbandry, in that it produces fish for the sea as well as fish. Many fish culture programmes are conducted not to raise food, but to raise fish that will be stocked in lakes, rivers, and oceans to help sustain or re-establish fisheries. These programmes are successful only if survival of the stocked fish is large enough to improve the creel or if the stocked fish spawn and increase the size of the next year-class. Unfortunately, many stocking programmes have not been successful, and a major reason for this is improper brood stock management at the hatchery. Brood stock management plans that make fish farming economically successful are plans that make stocking programmes unsuccessful. Much of the blame can be attributed to the small numbers of fish that are spawned every year which lead to unwanted levels of inbreeding that will, in turn, lower survival.

Even when fish are raised for food, the inbreeding produced by small populations can cause problems. If unwanted inbreeding occurs, it can reach levels that cause growth rate or other production phenotypes to decrease. This means yields will decline unless management is intensified; in other words, the environmental aspects of management must be increased to counteract genetic mismanagement. Furthermore, small population sizes produce genetic drift, which is random changes in gene frequency. The ultimate effect of genetic drift, particularly in small populations, is the loss of alleles. The loss of genetic variation can make future selective breeding programmes ineffective. Selection improves productivity by exploiting a population's genetic variance; genetic drift decreases a population's genetic variance.

Inbreeding can occur when a farmer conducts a selective breeding programme; in fact, it is inevitable. Selection is a breeding programme that improves a population by allowing only superior animals to produce offspring. When this happens, the size of the breeding population is reduced. When only the best are allowed to mate and the only criterion for mating is a certain phenotype, relatives are often mated. This causes unwanted inbreeding. While inbreeding will be produced during selection and while it can cause some problems, it is of secondary importance. The major genetic goal during this type of breeding programme is to improve the population by selection. However, inbreeding should be moderated when conducting a selective breeding programme, because there is no point in conducting selection simply to counteract inbreeding depression.

Consequently, the loudest and most persistent message that comes from geneticists and extension agents is: avoid inbreeding and do what it takes to prevent it from occurring. And that is the major thrust of this manual. Most of the manual describes what inbreeding is, what causes inbreeding, how it works genetically, how it is measured, and what can be done to prevent it from occurring. These subjects are developed in order to compile a list of recommendations that farmers and hatchery managers can incorporate into yearly work plans to better manage their populations and prevent inbreeding from ruining the population genetically. Since the problems associated with genetic drift and inbreeding are linked, part of this manual is devoted to this topic and to management techniques that can be used to minimise the effects of genetic drift.

A manual simply stating that inbreeding is bad and should be avoided would be only partially useful. Since inbreeding is an important breeding programme, one that has been used to produce superior animals and plants, there is no reason why aquaculturists cannot use inbreeding to produce superior stocks of fish. Consequently, one chapter is devoted to the ways inbreeding can be used to improve growth and other phenotypes and to the development of regular inbreeding programmes that can be used to produce inbred lines which can be used in selective breeding and crossbreeding programmes.

Few farmers will want to or be able to use inbreeding programmes. The decision to conduct any breeding programme is one that must be made for each farmer or each fry/fingerling production centre on a case-by-case basis. Breeding programmes are expensive and time consuming. They require a certain level of sophistication, because good record keeping is a requirement. This is especially true for inbreeding programmes. Inbreeding is the mating of relatives, and you cannot mate relatives if you cannot follow pedigrees. Also, these programmes require facilities. Ponds that are being used for inbreeding programmes are ponds that cannot be used to raise food. Finally, breeding programmes usually do not produce immediate improvements.

Improvements are usually not seen for at least one growing season; when inbreeding is used, there is often a 2- to 3-generation lag before improvements are noted. Consequently, a farmer must be able to incorporate long-term planning into his farm management programme, and he must be patient. As a result, within a region, only a small percentage of farmers or fingerling production centres should or will ever conduct inbreeding programmes. But even if few farmers or hatchery managers will conduct inbreeding programmes, an understanding of how inbreeding can be used to improve productivity will enable the rest to better understand what inbreeding is; how it occurs; what is does; and most importantly, how it can be controlled.

Genetics

An all-encompassing review of genetics. It is simply an overview of selected topics that will help those who do not have a good background in genetics to understand the genetics of inbreeding and thus understand how inbreeding affects production phenotypes, such as weight and fecundity. This background information will also help explain how inbreeding can be used to improve a population and how a population can be managed to minimise the effects of unwanted inbreeding.

Those who are not interested in this background information or those who already understand it can skip this chapter and go to: which explain how inbreeding values are calculated; which explains genetic drift; which describes how inbreeding can be used to improve a population; which discuss ways and techniques that can be used to manage a population in order to prevent unwanted inbreeding from causing inbreeding depression and to prevent genetic drift from decreasing genetic variance.

Genes and Chromosomes

Genes are located on structures called "chromosomes," which are located in the nucleus of every cell. Although there are some exceptions, chromosomes typically occur as pairs in animals, and for practical fish culture management, fish can be considered to be diploids (chromosomes occur in pairs). Some rare species of fish are triploids (chromosomes occur in groups of three), but none are aquacultured species. Some species of fish are tetraploids (chromosomes occur in sets of four), notably the salmonids and catostomids; however, for practical breeding work these species behave as if they were diploids, so they can be considered to be diploids.

Because chromosomes occur in pairs, each gene also occurs as a pair. There are some exceptions in fish with morphologically distinct sex chromosomes, but few fish have such chromosomes; consequently, for practical fish breeding work, genes can be considered to occur in pairs. One chromosome of each pair comes from a fish's mother, while the other comes from its father; this means one gene from each pair comes from the mother, while the other comes from the father. The two chromosomes that form a pair are called "homologues."

Meiosis

Meiosis is the process during which the primary gametocytes develop into eggs or sperm. A number of important cellular changes occur during meiosis, but for a geneticist, the important ones are those that affect the genes and chromosomes.

Meiosis consists of two cell divisions. The first meiotic division is called the "reduction division," and the second is called the "equational division." Prior to the reduction division, the homologues that form each chromosome pair replicate. The replicated homologues do not separate; they are joined together at the centromere, and the joined replicated homologues are called "sister chromatids".

The replication of the chromosomes is usually uneventful in that they replicate themselves perfectly. However, mistakes in the replication process occur. These mistakes are called "mutations." The mutation rate for each gene is low, ranging from one in 10,000 replications to one in 100,000 replications.

When a mutation occurs, the erroneous copy of the gene is often capable of producing an alternate version of the original phenotype. Most mutations are harmful, but some mutations create new versions of genes, and these new variants produce new or improved phenotypes that increase survival; this can lead to the evolution of new species. If replication were perfect, life would not have evolved beyond the single-celled state. Mutations also help explain some of the consequences of inbreeding.

During the early phase of the reduction division, the replicated homologues that form each chromosome pair come together. Because each homologue is composed of a pair of sister chromatids, each bundle is composed of four chromosomes called a "tetrad."

The chromosomes that form a tetrad typically twist around each other and break. When the chromosomes rejoin, pieces can be exchanged between homologues. When this happens, pieces of chromosomes, along with the genes that are located along those pieces, are transferred from one homologue to the other. This process is called "crossing over". Crossing over is one of the most important biological processes, because it greatly increases genetic variability.

In inbreeding, crossing over is important because it can affect the amount of inbreeding produced during one form of chromosomal manipulation. Crossing over tends to occur at particular locations along each chromosome. Each gene has a given cross over frequency. Genes located near the centromere cross over less frequently than those further away. Gene that are located near each other on a chromosome cross over as a unit and are called "linked"; the closer they are, the more tightly linked they are. Consequently, crossing over causes actual levels of inbreeding to deviate from predicted levels during meiotic gynogenesis.

During reduction division, the chromosome number is reduced from the diploid (paired) state to the haploid (unpaired) state. The homologues that form each chromosome pair separate and are parcelled into the two daughter cells (for males, the two secondary spermatocytes; for females, the secondary oocyte and first polar body). When the chromosomes divide, they are not parcelled into maternal and paternal sets. The direction in which each replicated homologue (sister chromatids) of each chromosome pair goes is random and independent of that for all other chromosomes. This process is called "independent assortment." Although each secondary gametocyte contains a haploid set of chromosomes, each chromosome is duplicated.

The second meiotic division is called the "equational division." During the equational division, the replicated homologues (sister chromatids) in each secondary gametocyte separate and are parcelled into either two sperm cells or into the egg and second polar body. This produces haploid gametes; each gamete contains a single chromosome from each pair, which means that it contains a single copy of each gene.

Phenotype and Genotype

Each gene contains the chemical blueprint for the production of a protein. This protein either forms or helps produce a specific phenotype (also called "trait" or "character"), such as body colour or weight. A phenotype is the physical or chemical expression of what a gene produces; it is either observed and described (qualitative phenotype) or it is measured (quantitative phenotype).

A gene can occur in one or more forms. Alternate forms of a gene are called "alleles." In a population, a gene can have one to perhaps a dozen alleles. If there is only one allele in the population, the gene is said to be "monomorphic." If there are two or more alleles at a locus (locus = gene; plural is loci), the gene is said to be "polymorphic." Monomorphic genes are not very interesting, because there is no genetic or phenotypic variance associated with that locus. On the other hand, polymorphic genes are of great interest to a geneticist. When more than one allele exists at a locus, there is genetic variance, which produces phenotypic variance; and this can be exploited by breeding programmes.

The genetic make-up of each fish is called its "genotype." Because chromosomes occur as pairs, each gene occurs as a pair; this means the genotype is a paired entity. If the copy of the gene that a fish inherits from its father is the same as the copy that it inherits from its mother, that gene is an identical pair of alleles, and the fish is said to be "homozygous" at that locus. If the copy of the gene that a fish inherits from its father is different from the copy that it inherits from its mother, the gene is a non-identical pair of alleles, and that fish is said to be "heterozygous" at that locus. Consequently, every gene can exist in one of two genotypic states: homozygous (identical pair of alleles) and heterozygous (non-identical pair of alleles).

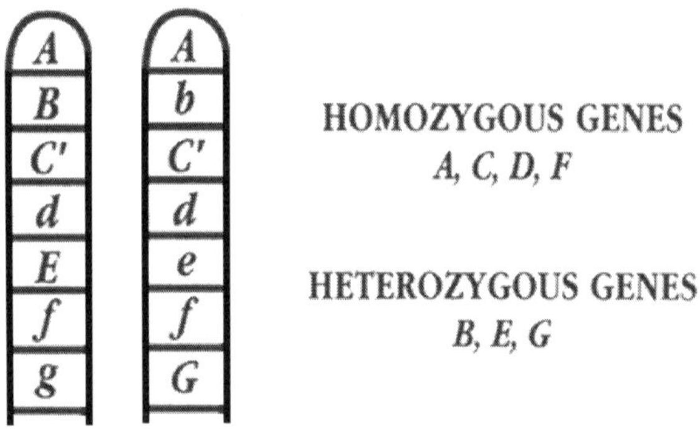

Figure: A small section of a pair of chromosomes, showing seven genes. Four of the genes exist in the homozygous state (A, C, D, and F), while three exist in the heterozygous state (B, E, and G).

How homozygous and heterozygous fish are produced. In this example, both parents were heterozygotes (Aa), so each produced two types of gametes: half the eggs and half the sperm carried an A allele, while the other half carried an a allele. Some progeny were homozygous (AA or aa) because they inherited the same allele from both parents.

Some progeny were heterozygotes (Aa) because they inherited a different allele from each parent.

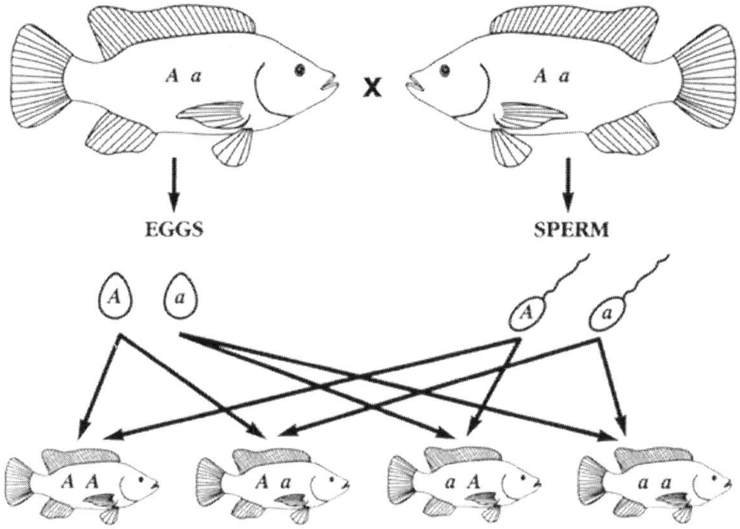

Figure: Progeny With *Aa* And *Aa* Genotypes Are Homozygous; They Inherited Identical Alleles From Both Parents

The reason why it is important to make a distinction between the homozygous and heterozygous states, is they often produce different phenotypes. The number of phenotypes produced by a gene depends on the mode of gene action and the number of alleles. A fish's genotype produces its phenotype. Consequently, if a breeder wants to conduct a breeding programme to control or alter phenotypic frequencies (using inbreeding or preventing it are both breeding programmes), he should understand how the genotype produces the phenotype.

Genetics of Qualitative Phenotypes

Breeders divide phenotypes into qualitative (descriptive) phenotypes and quantitative (measured) phenotypes. Qualitative phenotypes are those that are described, such as body colour. Because these phenotypes are descriptive, individuals segregate themselves into descriptive categories: pink vs. normally pigmented; gold vs. bronze vs. black; normal vs. deformed.

In general, qualitative phenotypes are controlled by one or two genes. The alternate forms of a phenotype (pink vs. normally pigmented or normal vs. deformed) are produced by the alternate forms of a gene (the alleles).

Most qualitative phenotypes that have been described genetically in fish are controlled by a single autosomal gene with two alleles (autosomes are the pairs of chromosomes that males and females share in common; the pairs of chromosomes that are not the pair of sex chromosomes). Because of that, this discussion will concentrate on these phenotypes, especially since this type of inheritance helps explain how inbreeding works.

In general, genes express themselves in either an additive or a non-additive manner. When genes express themselves in a non-additive manner, one allele is dominant, while the other is recessive. The terms "dominant" and "recessive" carry no value judgement; the terms simply refer to the way the alleles produce their respective phenotypes.

There are two types of dominant alleles. When the dominant allele exhibits complete dominance, it always produces its phenotype, while the recessive allele can produce its phenotype only in the absence of the dominant allele. Thus, a gene that produces two phenotypes via complete dominance (it has a completely dominant allele and a recessive allele) has three genotypes, but the genotypes produce only two phenotypes:

Genotype	*Phenotype*
Homozygous dominant	Dominant
Heterozygous	Dominant
Homozygous recessive	Rece
ssive	

This type of gene action plays an important role in the genetics of inbreeding, and it helps explain inbreeding depression. This topic will be discussed later in this chapter.

A second type of dominance occurs when the dominant allele is incompletely dominant; i.e., the dominant allele always produces its phenotype, but it is unable to completely suppress the recessive allele in the heterozygous state. When this happens, a gene with incomplete dominant gene action (it has an incompletely dominant allele and a recessive allele) has three genotypes, and each genotype produces a unique phenotype:

Genotype	*Phenotype*
Homozygous dominant	Dominant
Heterozygous	Near-dominant
Homozygous recessive	Recessive

The heterozygous genotype produces a phenotype that resembles but is slightly different from the dominant phenotype.

When the mode of gene action is additive, neither allele is dominant, and each allele always produces its phenotype in a unidirectional step-wise manner. This means the heterozygous phenotype is intermediate between the two homozygous phenotypes. When this occurs, a gene with additive gene action produces a unique phenotype for each genotype:

Genotype	*Phenotype*
Homozygous allele 1	Homozygous phenotype 1
Heterozygous	Heterozygous
Homozygous allele 2	Homozygous phenotype 2

Only one additive gene has been discovered in fish; it controls golden, palomino, and normal body colours in rainbow trout, Oncorhynchus mykiss. All other qualitative phenotypes that are produced by a single autosomal gene are controlled either by complete dominance or incomplete dominance.

Genetics of Quantitative Phenotypes

Quantitative phenotypes are those that are measured, such as length, weight, and fecundity. The important production phenotypes are quantitative ones, although some qualitative phenotypes are quite important, and can greatly increase the value of a crop. Because quantitative phenotypes are measured, each phenotype is a single category, such as length. Individuals do not get segregated into alternate phenotypic categories, such as long vs. short; instead, they are distributed along a continuum, and differences among individuals are determined by the unit of measure that is used to assess the phenotype: millimetres, grams, etc.

Because each fish's phenotype is determined by a measurement, quantitative phenotypes form what are called "continuous distributions," which are described by the population's mean and the distribution about the mean (variance and standard deviation). In a population, these phenotypes form what are called "normal" or "bell-shaped" distributions.

The reason why quantitative phenotypes exhibit continuous distributions and why individuals do not get segregated into descriptive categories is that quantitative phenotypes are far more complicated genetically than qualitative phenotypes. Each quantitative phenotype is controlled by dozens to hundreds of genes, each of which makes a small contribution to the production of the phenotype. The exact number of genes is usually never known. Additionally, each phenotype is strongly influenced by the environment (e.g., date of birth, access to food, age of mother, etc.), and the influence varies both from family to family and from individual to individual. The simultaneous action of many genes and the environmental effects creates single phenotypic categories (e.g., length), in which the only way to describe an individual is to measure it.

Because each quantitative phenotype is controlled by numerous genes, as well as environmental variables, the only way to work with these phenotypes is to analyze the phenotypic variance that exists in the population and to divide the phenotypic variance into its component parts. Phenotypic variance (V_P) is the sum of the genetic variance (V_G), environmental variance (V_E), and genetic-environmental interaction variance ($V_{G\text{-}E}$) components:

$$V_P = V_G + V_E + V_{G\text{-}E}$$

When conducting a breeding programme, a geneticist tries to exploit V_G. Three distinct types of genetic variance combine to make V_G, and it is important to know what they are, because different breeding programmes are needed to exploit each type. Genetic variance is the sum of additive genetic variance (V_A), dominance genetic variance (V_D), and epistatic genetic variance (V_I):

$$V_G = V_A + V_D + V_I$$

These components of genetic variance do not refer to additive, dominance, and epistatic gene action; they refer to specific components of phenotypic variance that are produced by the entire genome, not that produced by one or two genes.

The major components are V_A and V_D. Epistatic genetic variance is usually considered to be unimportant, because it is difficult to exploit and improvements that occur by exploiting V_I plateau quickly.

Additive genetic variance is the component that is due to the additive effect of all the fish's alleles taken independently; i.e., the sum of the effects that each allele makes to the production of the phenotype. Additive genetic variance is the most important component of V_P, and the percentage of V_P that is controlled by V_A is called "heritability" (h^2):

$$h^2 = V_A/V_P$$

Additive genetic variance is the genetic component that can be exploited by selective breeding programmes. In general, phenotypes with h^2's $\gtrsim 0.25$ can be improved efficiently by individual (mass) selection; when h^2's are $\lesssim 0.15$, family selection is needed to improve a population. When h^2 is 0.10, family selection is often ineffective, although some gains can be made when inbreeding is combined with family selection.

Dominance genetic variance is the other major component of V_G. Dominance genetic variance is the component that is due to the sum of each interaction that exists between the two alleles at each locus. Because V_D is produced by the interaction of the alleles at a locus, V_D cannot be inherited from either parent. Dominance genetic variance is a function of the diploid state (a function of the paired gene), and offspring inherit alleles that exist in the haploid state from each parent. Dominance genetic variance is created at fertilization when the haploid set of chromosomes from the mother pairs with the haploid set of chromosomes from the father. When this occurs, each gene exits in the paired state, and V_D is created. Consequently, V_D is destroyed by meiosis, and it is recreated anew and indifferent combinations at fertilization.

The breeding programme needed to exploit V_D is crossbreeding. When h^2 is 0.15, crossbreeding is often prescribed to exploit V_G.

Inbreeding

Inbreeding is the mating of relatives. Genetically, all inbreeding does is increase homozygosity in the offspring (this means inbreeding also decreases heterozygosity in the offspring by an equal amount; in this manual, inbreeding will be defined and discussed as it relates to homozygosity). The increase in homozygosity occurs because related fish share alleles though one or more common ancestors; i.e., the parents may carry a copy of an allele that both inherited from a common ancestor. When relatives mate, the alleles that they share because of their common ancestor(s) can be paired in their offspring. This produces offspring that are more likely to be homozygous at one or more loci.

The mating of unrelated fish also produces offspring that have homozygous genes. Additionally, an inbred fish looks the same as one with no inbreeding; there is no distinguishing mark that separates fish into inbred vs. non-inbred categories. So why is a distinction made? How are they different genetically?

The answer to these questions are: The two types of homozygosity are identical; there is no genetic difference. The only difference between inbred and non-inbred fish is that the homozygosity of each is created in different ways, but this distinction is most important. Inbred fish are homozygous because they have genes where the two alleles are identical by descent; non-inbred fish are homozygous because they have genes where the two alleles are identical because they are alike in kind. Inbred fish are homozygous at a locus because they inherited identical copies of an allele from both parents that the parents, in turn, inherited from a common ancestor. Non-inbred fish are homozygous at a locus because they just happened to inherit an identical pair of alleles from their parents. This difference in the type of homozygosity is not physical or chemical. The only difference is the paths that the alleles took before they were paired to create the homozygosity; i.e., how the alleles were inherited.

Since there is no difference, why do geneticists make a distinction between the two types of homozygosity? Why is inbreeding of any concern? The reason is that related fish are more alike genetically than unrelated fish. This means that when relatives mate, they produce offspring that have more homozygous loci than is the case when unrelated fish mate.

The mating of relatives produces offspring that tend to be more homozygous than the population average. This increase in homozygosity is called "inbreeding," and the coefficient of inbreeding (F) is a measure of how much more homozygous a fish is than the population average. The coefficient of inbreeding does not measure how many homozygous loci the fish has; it simply quantifies the percent increase in homozygosity. On a gene-by-gene basis, F is the probability that the two alleles will be identical by descent.

Since F measures the percent increase in homozygosity, the same level of inbreeding can produce different amounts of homozygosity in different populations, depending on the level of homozygosity that existed before inbreeding began. The coefficient of inbreeding merely quantifies the increase in homozygosity that has occurred over a specific time interval (number of generations) due to identity by descent (due to the mating of relatives). Additionally, F is based on probability, so a given value of F is an average value; consequently, a given value of F means more homozygosity will have been produced in some fish, while less will have been produced in others.

The amount of inbreeding that is produced when relatives mate depends on the closeness of the relationship between the parents. Close relatives (e.g., brothers and sisters; parents and offspring) share many alleles inherited from common ancestors, so when they mate, they produce offspring with high levels of inbreeding. Distantly related individuals (e.g., fourth cousins; fifth cousins) share comparatively few alleles from a common ancestor, so when they mate, they produce offspring with comparatively little inbreeding. The way inbreeding is measured is explained in Chapters 3 and 4, and the inbreeding that is produced when various relatives mate is explained.

Because inbreeding increases homozygosity, it changes genotypic frequencies, increasing the percentage of homozygous genotypes, while shrinking the percentage of heterozygous genotypes. If regular systems of inbreeding are conducted, the breeding programme creates a series of inbred families which do not interbreed. This subdivides a population into numerous lines, which further increases genotypic variance.

While inbreeding changes genotypic frequency, it does not change gene frequency. Selection, genetic drift, migration, and mutation are the evolutionary and breeding forces that change gene frequency. Inbreeding itself does not alter gene frequency, but by altering genotypic frequency inbreeding can, theoretically, accelerate selection.

During an inbreeding programme, genotypic frequencies can be drastically altered if relatively few inbred families are maintained (a type of selection). When this occurs, gene frequencies will also change.

Chapter 3

Uses of Inbreeding

Inbreeding is a breeding programme that can be used to produce superior animal and plant brood stock, and it can also be used to produce genetically improved animals and plants for grow-out. Inbreeding is the breeding programme that is often used to create new breeds or varieties that breed true for "type"; i.e., a particular body conformation or set of qualitative phenotypes. Linebreeding is a form of inbreeding that is used to increase an outstanding animal's contribution to a population. The use of herd bulls is a form of inbreeding that is used to quickly improve a population. Inbreeding can be used as a type of progeny testing to create defect-free animals. Inbreeding can be used to improve response to selection when h^2 is small by combining inbreeding with between-family selection.

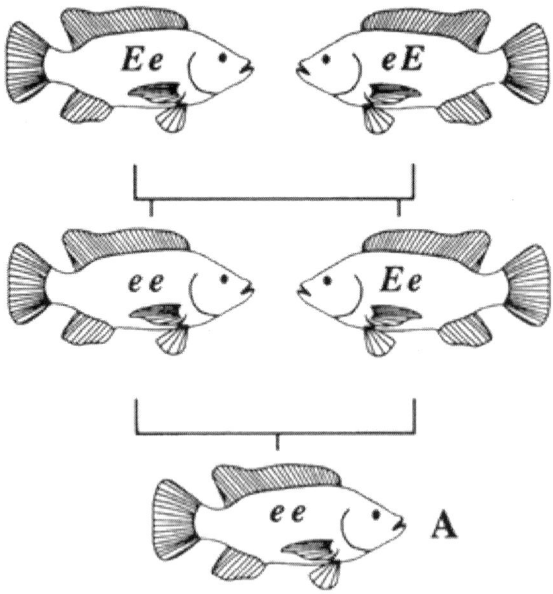

Figure: Fish A Is Homozygous By Descent; Fish A Is Inbred

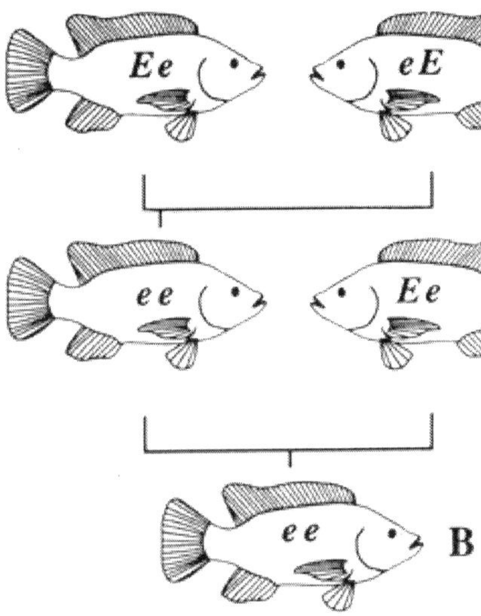

Figure: Fish B Is Homozygous In Kind; Fish B Is Not Inbred

Inbreeding Depression

As inbreeding increases, it often causes a decrease in productivity which is termed "inbreeding depression." Inbreeding depression is a decrease in growth rate, fecundity, etc. that is observed in the inbred group when it is compared to a control population where there is no inbreeding. The severity of inbreeding depression depends on the level of inbreeding, the phenotype in question, and the population. Inbreeding depression is what gives inbreeding its bad reputation.

There are several explanations for why inbreeding depression occurs, and all contribute to inbreeding depression. The first two are qualitative genetic explanations.

The most common explanation is that inbreeding depression occurs because of the pairing and expression of detrimental recessive alleles. Although the terms "dominant" and "recessive" do not mean that one allele is good and the other undesirable, most of the alleles that produce abnormal phenotypes or that lower viability are recessive. Even though mutation rates are usually only one in 10,000 to one in 100,000 replications per gene, because each fish can produce many gametes (up to 500,000,000 sperm per spawn) and because each fish has tens of thousands of genes in its genome, each fish produces dozens to hundreds of gametes that contain a mutant copy of one or more genes. Most of these mutant alleles are recessive. Since recessive alleles can be expressed only when a fish is homozygous, these mutations tend to accumulate in a population. Consequently, many fish carry "hidden" copies of these mutant alleles in the heterozygous state.

Mutations do not have to be bad; some increase fitness or lead to the evolution of new species by creating new or improved phenotypes. But most mutations lower fitness, because they cause random changes in a phenotype that works and that works well as a result of natural selection. Many mutant alleles produce phenotypes that are so abnormal that they either cause death or reduce viability severely. Others cause only small reductions in viability. Each fish carries a number of these detrimental recessive alleles, and inbreeding uncovers them. When relatives mate, they produce offspring with an increased level of homozygosity; consequently, some of the detrimental recessive alleles that the parents carry in the unexpressed heterozygous state are paired and expressed in the offspring. Detrimental recessive alleles can also be paired and expressed when unrelated fish mate. The difference is one of magnitude, and that is what causes inbreeding depression.

It has been estimated that each individual carries dozens of mutant alleles that lower viability. Most of these mutant alleles will produce only a small reduction in fitness, but several will produce phenotypes so abnormal that they are lethal or cause premature death. There are many of these unexploded genetic bombs in the gene pool. If two unrelated individuals mate, there is a good chance that they will not carry the same genetic dynamite. However, when relatives mate, there is a good chance that both parents will carry the same genetic time bombs, because they inherited them from a common ancestor. Even though the probability of producing a defective offspring is the same for each recessive allele (25% if both parents are heterozygotes), if both parents carry a large number of identical defective recessive alleles, the odds of producing defective offspring increase dramatically. The more closely related the parents, the more alike they are genetically, so it is more likely that defective offspring will be created.

Even if two parents do not carry any lethal recessive alleles, they can produce inbred offspring which exhibit inbreeding depression if the parents carry a number of recessive alleles that reduce viability, growth, or other production phenotypes. For example, if two relatives carry 10 recessive alleles, each of which depresses growth by 1%, the pairing and expression of these alleles will depress growth in the inbred offspring. Some offspring produced by these parents will not be homozygous for any of these genes, but most will be homozygous for one or more of these recessive alleles, and growth will be depressed from 1–10% in these offspring. A second explanation for inbreeding depression is that by decreasing heterozygosity, inbreeding reduces what is called "overdominance." Overdominance occurs when the heterozygous genotype produces a phenotype that is superior to the two homozygous ones. Because inbreeding increases homozygosity, it decreases the number of heterozygous loci, which means that it reduces overdominance. The best known example of overdominance occurs in humans. In many populations, the gene which produces hemoglobin exists in two allelic forms: one produces normal hemoglobin, while the other produces sickle-cell hemoglobin. Individuals who are homozygous normal are susceptible to malaria and get quite ill and can die when infected. Individuals who are homozygous for sickle-cell get sickle-cell anemia and are sickly and often die. The heterozygotes are superior, even though they are slightly anemic, because they are resistant to malaria.

The final explanation for inbreeding depression is a quantitative genetic one. If a quantitative phenotype were controlled solely by V_A, inbreeding would not effect the phenotype, because the types of homozygosity that would be created at each locus would balance over the genome: some homozygous combinations would cause a slight decrease in the phenotypic mean, while other would cause a slight increase. Since quantitative phenotypes are controlled by hundreds of genes, and since the homozygosity that is created by inbreeding tends to be random, if a quantitative phenotype were solely controlled by V_A, inbreeding-produced homozygosity would neutralize itself, in terms of phenotypic expression. However, few if any quantitative phenotypes are controlled solely by V_A. When V_A controls just 30% of V_P for a particular phenotype, it produces a $h^2 = 0.3$, which is considered to be large.

Since inbreeding increases homozygosity, it also decreases heterozygosity. By doing this, inbreeding affects the interactions between alleles at each locus, which means that it affects V_D, and dominance effects can play a major role in the production of many quantitative phenotypes. When h^2 is small (<0.15), as it often is for viability and fecundity, dominance effects can be quite significant. The creation of excess homozygosity tends to disrupt dominance effects, which tend to be more unidirectional than the additive effects. Inbreeding can be considered to be the opposite of crossbreeding, which is used to exploit V_D. Consequently, if crossbreeding maximises exploitation of V_D, inbreeding minimises exploitation, which means that inbreeding makes phenotypes controlled by V_D worse (inbreeding decreases phenotypic mean). This genetic explanation makes sense, because inbred lines are often used in crossbreeding programmes. The creation of inbred lines is done to produce opposite types of homozygosity that will be maximised when hybrids are created, which is how V_D is exploited. This explanation is further supported by the fact that crossbreeding, the breeding programme that is used to exploit V_D, can also be used to eliminate inbreeding and inbreeding depression.

Inbreeding Studies in Fish

Unfortunately, there have been very few inbreeding studies with fish, so there is little specific information about the effects of inbreeding in aquacultured populations. In fact, inbreeding is the least studied aspect of fish genetics. Most of the studies that have been conducted have been with salmonids, particularly rainbow trout. Most of the studies are singular experiments that were limited in scope, and many investigated only a single level of inbreeding-usually 25%. Furthermore, most studies looked at the effects of inbreeding in the range of 25–60%. No study has looked at the effects of inbreeding <12.5%. Such studies are needed, because inbreeding experiments with other animals have shown that mild levels of inbreeding can be beneficial. Because of this, we know relatively little about the effects of inbreeding in fish culture. The studies have shown a few preliminary trends, but we do not know how different levels of inbreeding affect various phenotypes in all important cultured species of fish. We do not know what levels of inbreeding are beneficial and what levels are harmful. Finally, no inbreeding studies have been conducted on some important aquacultured species. In general, inbreeding studies with fish have shown that inbreeding decreased production phenotypes such as growth rate, fecundity, and survival, while increasing the number of deformed offspring.

The most complete study was a long-term evaluation of the effects of six levels of inbreeding on rainbow trout. The results show that the mildest level of inbreeding that was produced (12.5%; this is equal to one generation of half-sib matings) adversely affected percent hatch and survival, but it increased 77- to 150-day weight. Levels of inbreeding e"25% severely depressed fecundity, growth, and survival (inbreeding of 25.0%, 37.5%, 50.0%, and 59.4% are produced by one, two, three, and four generations of brother-sister matings). It is not known if these results apply to all species or to all populations of rainbow trout. Studies that have been conducted with other species of fish showed that inbreeding usually decreased means for production phenotypes, although the studies have produced mixed results: for example, inbreeding reduced the return rate of Atlantic salmon, Salmo salar;

reduced growth and viability of Mozambique tilapia, Oreochromis mossambicus; increased growth of channel catfish, Ictalurus punctatus, in one study but reduced it in another. The only unambiguous result was that inbreeding produced some abnormal offspring. It has been suggested that the breeding programmes used by most fish farmers will produce inbreeding of 3–5% per generation. If this occurs, inbreeding depression could begin to affect productivity and profits after only 3 to 5 generations.

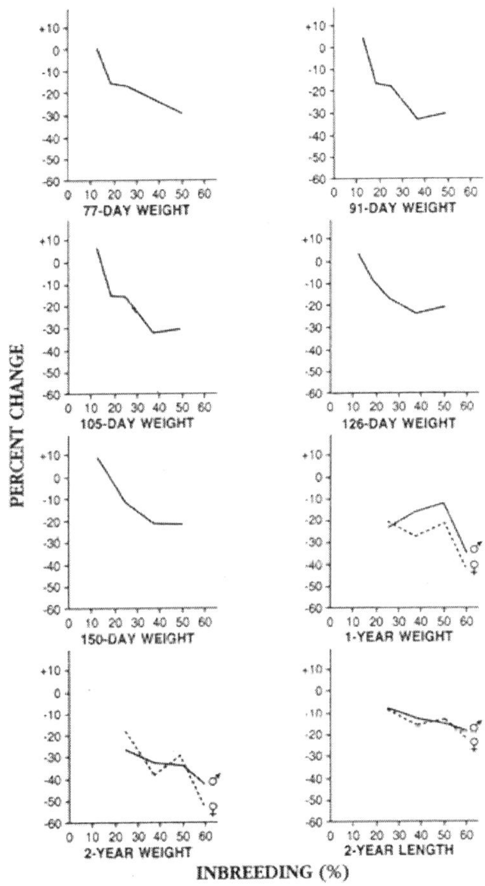

Figure: The effects of inbreeding on length and weight in rainbow trout. The results are presented as percent change when the inbred group was compared to a control group (inbreeding of 0%); a negative number means the inbred group was worse than the control and exhibited inbreeding depression. One generation of brother-sister mating produces inbreeding of 25%; two generations produce inbreeding of 37.5%; three generations produce inbreeding of 50%; and four generations produce inbreeding of 59.4%.

Data used to construct this figure are from: Kincaid, H.L. 1976. Effects of inbreeding on rainbow trout populations. Transactions of the American Fisheries Society 105:273–280; Kincaid, H.L. 1983. Inbreeding in fish populations used for aquaculture. Aquaculture 33:215–227.

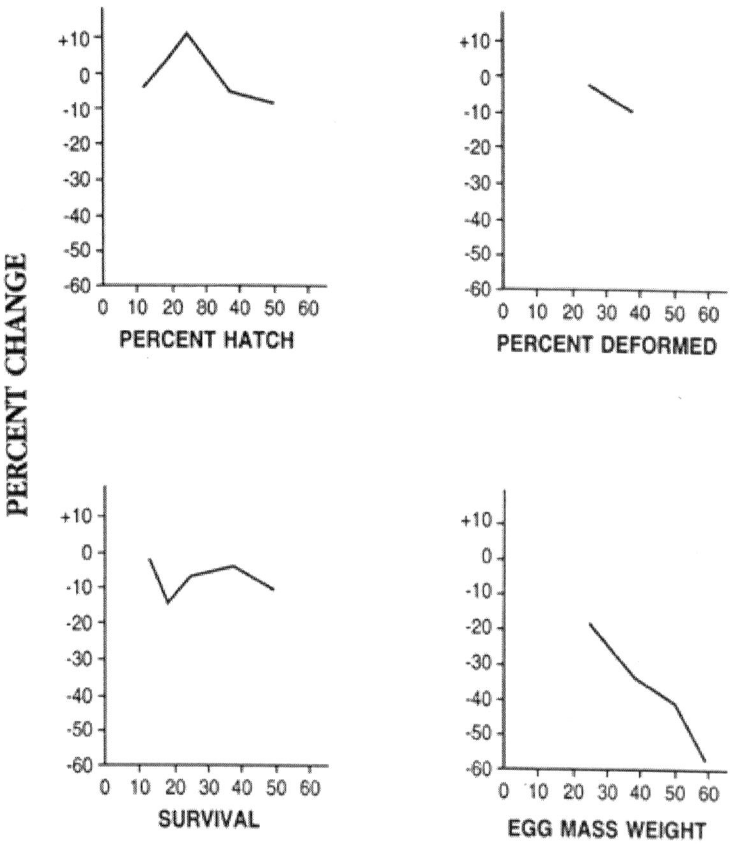

Figure: Effect of inbreeding on hatching rate, number of deformed offspring, survival, and fecundity (egg mass weight) in rainbow trout. The results are presented as percent change when the inbred group was compared to a control group (inbreeding of 0%); a negative number means the inbred group was worse than the control and exhibited inbreeding depression.

Calculating Individual Inbreeding Values

Calculating individual inbreeding values often requires information that a farmer or hatchery manager does not have-pedigrees. Those who culture fish usually do not know a fish's mother or father. Even if this information is known, it is unlikely that grandparents or cousins can be identified. This information is routinely gathered in livestock husbandry, and family pedigrees are recorded on forms or are entered into computer data bases.

It is far easier to record this information for cattle, pigs, sheep, and goats, because these animals are large, live on land, and can be branded or tagged with little effort at birth. This enables those who work with the marked livestock to identify each individual, to record matings and births, and thus to create pedigrees. Few aquaculturists record such information with fish, because fish are difficult to mark, and marking techniques often kill a large percentage of the fish. Additionally, while this information has been considered important for centuries with livestock, it is still considered to be of little value by most aquaculturists.

Even though it is impossible for most farmers or research scientists to determine individual inbreeding values, it is important to know how these values are calculated and to learn the techniques that are used. An understanding of the methods and protocols needed to determine individual inbreeding values explains how inbreeding is created. It helps explain how an individual can become homozygous by descent, which is what occurs when relatives mate and produce offspring. Learning how to trace family trees and how to calculate individual inbreeding values demonstrates how an ancestral allele can be inherited, both maternally and paternally, from a common ancestor. An understanding of these techniques also shows how inbreeding can be prevented, demonstrates how inbreeding can be reduced to zero, and helps lay the groundwork for the next chapter, the determination of average inbreeding values in hatchery populations.

Two methods can be used to determine individual inbreeding values: path analysis and covariance analysis. Both methods produce the same answer, but some feel more comfortable with one than the other. Both methods use simple arithmetic. Path analysis is quicker and requires fewer steps, but it is easier to make a mistake using this technique. Covariance analysis can take far more time if the pedigree is complicated, but once complete, the information generated by covariance analysis can be used to quickly predict inbreeding values that will be produced from any possible mating between individuals in the pedigree.

Creating a Pedigree

The first step in determining an individual fish's inbreeding is to create its pedigree. A pedigree is a family tree. It lists an individual's brothers and sisters, parents, aunts and uncles, grandparents, great-grandparents, nephews and nieces, and cousins. The information is usually streamlined; often, only direct ancestors are included; and the family tree is often traced back four or fewer generations, only going back to an individual's great-grandparents or occasionally including the great-great-grandparents.

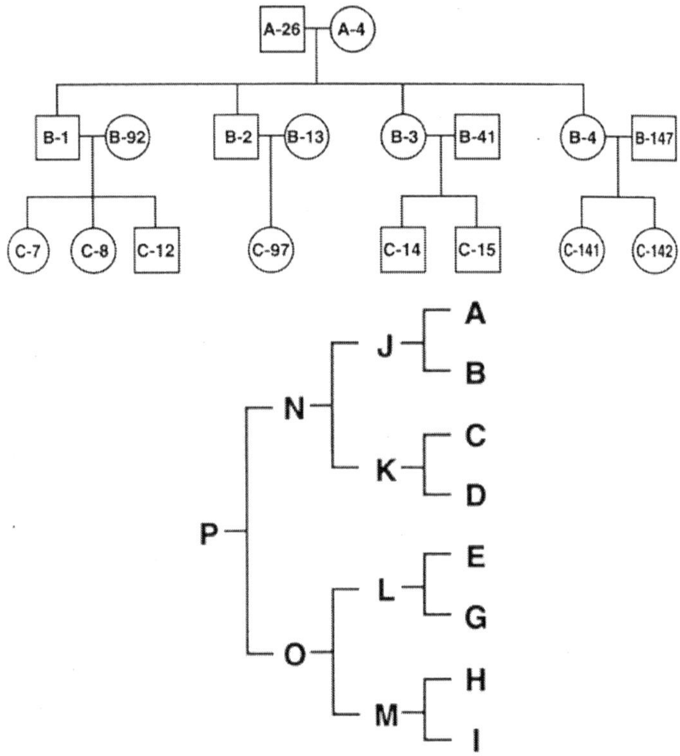

Figure: Examples of two pedigrees. The top pedigree uses squares to represent males and circles to represent females. Fish name is inside each square or circle. The other pedigree simply uses letter symbols to represent fish. Information about the fish, such as birth date, weight, etc., would be kept in ledgers or on data sheets. When letters are used to designate individuals, the letter F is usually omitted, because F is the symbol for coefficient of inbreeding; this prevents confusion.

More generations can be listed, but with each additional ancestral generation, the genetic contribution of each fish in the oldest generation to the present generation decreases to the point where it becomes genetically meaningless. This is because an animal's genetic contribution to its descendants is halved each generation: Each parent contributes 50% of a individual's genes; each grandparent contributes

25%; each great-grandparent contributes 12.5%; and each great-great-grandparent contributes 6.25%. The contribution of each great-great-great-great-great-grandparent (seven ancestral generations) is only 0.78%. These values are determined by the number of direct ancestors an individual has in each ancestral generation: two parents, four grandparents, eight great-grandparents, and 16 great-great-grandparents. All fish have 128 great-great-great-great-great-grandparents, which means the contribution from each is so small (0.78%) that it is of no value in most breeding programmes.

Traditionally, males are represented by squares and females are represented by circles. Various symbols can be added to the squares and circles to convey phenotypic information, such as body colour, genetic diseases, etc., but this information is not needed to determine inbreeding. Other information that can be added to a pedigree are: names, birth and death dates, and weights at specific periods. When letters are used to designate individuals, the letter F is usually omitted, because F is the symbol for coefficient of inbreeding; its omission prevents confusion.

Path Analysis

In path analysis, a pedigree is converted to a path diagram. A path diagram differs from a pedigree in that arrows between parents and offspring replace the tier-step decent brackets that are used in a pedigree. Each arrow represents a gamete, which means that each arrowhead represents the path or way an individual received 50% of its genes, and the arrow shaft touches the parent that contributed those genes. Thus, an arrow shows how a fish received one homologue of each chromosome pair, which means it also shows how a fish received one allele of each allelic pair (each gene).

Each fish in a path diagram can be touched by zero, one, or two arrowheads. If a fish's parents are unknown, no arrowhead will touch it; if one parent is known, a single arrowhead will touch the fish; if both parents are known, the fish will be touched by two arrowheads. A fish cannot be touched by more than two arrowheads, because no fish can have more than two biological parents. If this occurs, you have made an error. (Eventually, various biotechnological manipulations will enable research scientists to create fish that have three or four parents; however, for everyday practical breeding work, this situation can be ignored.) Many arrows can leave an individual, if that fish produced a number of offspring that are listed in the family tree that is being analyzed.

The arrows, or paths, flow from the oldest to the youngest generation, and they show how the genes were transmitted from generation to generation. They do not always flow in a straight line and sometimes cross or go sideways, depending on the complexity of the family tree and the matings that have occurred. Because of this, it is crucial to make the path diagram large enough to house the arrows and to accurately place the arrows. If arrows are incorrectly drawn, inbreeding values generated by the path diagram will be incorrect. An individual's inbreeding value is calculated by determining all possible paths that the individual has with one or more common ancestors. A common ancestor is an individual that occurs on both sides of the family tree or pedigree; i.e., an individual that contributes genes through both the mother and the father. If just one common ancestor exists, the individual's parents are related and the fish is inbred. If there are no common ancestors, the inbreeding of that fish is zero. Obviously, if family trees are traced back far enough, all species of fish, like all humans, are related though a common ancestor. However, for practical breeding purposes, common ancestors that are five or six ancestral generations removed are of little importance in any breeding programme, so the genetic contribution of such distant relatives is usually ignored. In most western societies, the legal distance for consanguineous marriages is second cousins; i.e., it is illegal to marry a first cousin or someone more related than that, but is lawful to marry a second cousin or someone less related than that. Western societal norms figure that the inbreeding produced by second cousin marriages is insignificant and therefore acceptable, while that produced by first cousin marriages is not.

The following formula is used to determine individual inbreeding values from a path diagram:

$$F_X = \Sigma[(0.5)^N(1 + F_A)]$$

where: F_X = the inbreeding of an individual;

Σ = the symbol for "sum of " or "add";

N = the number of individuals in a path that is determined by tracing a path from one parent back to the common ancestor and forward from the common ancestor to the other parent; if more than one common ancestor exists, the term $(0.5)^N$ is repeated for each common ancestor; if more than one path exists between the individual and a common ancestor, the term $(0.5)^N$ is repeated for each unique path;

F_A = the inbreeding of the common ancestor.

If the inbreeding value of the common ancestor is zero or unknown, in which case you must assume that it is zero, the formula is simplified to:

$$F_X = \Sigma(0.5)^N$$

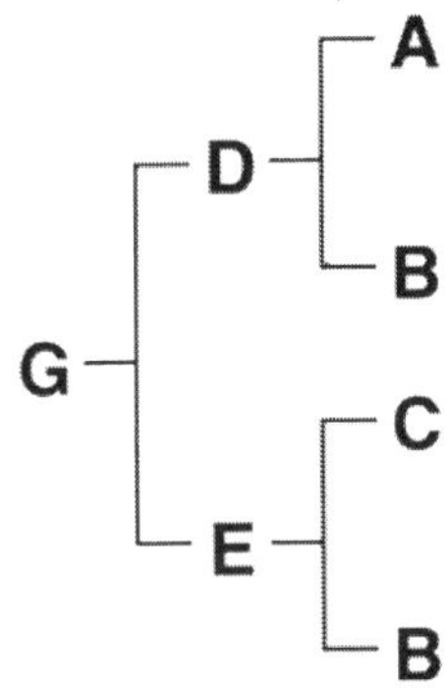

Figure: Pedigree

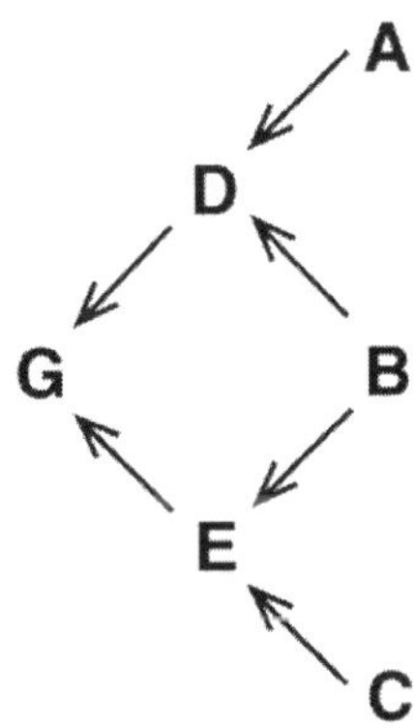

Figure: Path Diagram

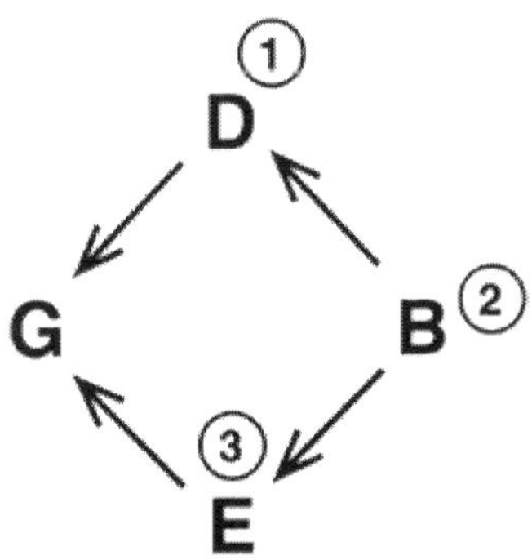

Figure: Path From G to Common Ancestor B

The formula may look like complicated math, but it is simple arithmetic that can be easily done with basic, inexpensive hand-held calculators; the term $(0.5)^N$ can be determined by using the "y^x" button.

Fish G is the only fish that has a common ancestor. An inspection of the pedigree shows that fish B is on both sides of fish G's family tree; this means that fish G is inbred. No other fish depicted in the pedigree has a common ancestor. To determine an individual's inbreeding, a path is traced from one parent back to the common ancestor and from the common ancestor up to the other parent. Consequently, fish G's inbreeding is determined by tracing the path that can be drawn from fish G's parents to fish B; i.e., from fish D (one parent) to fish B (the common ancestor) and from fish B to fish E (the other parent). Therefore, the path that is traced goes:

D-B-E

There are three individuals in this path, so N = 3. Since fish B (the common ancestor) is not inbred, the simplified formula can be used to calculate the inbreeding value of fish G:

$$F_G = (0.5)^3$$
$$F_G = 0.125$$

Thus, $F_G = 0.125$ or 12.5%. This means that fish G is expected to have 12.5% more homozygous loci than the average fish in the population. This also means that 12.5% of the loci that were heterozygous became homozygous as result of inbreeding. A more complicated pedigree and its path diagram. Fish J and fish N are inbred, because both have common ancestors. Fish A is the common ancestor of fish J. The inbreeding of fish J is determined by tracing a path to fish A through fish J's parents (fish H and fish I). The path is:

H-A-I

There are three individuals in the path, so N = 3. Since fish A (the common ancestor) is not inbred, we can use the simplified formula, and the inbreeding value for fish J is:

$$F_J = (0.5)^3$$
$$F_J = 0.125$$

Thus, $F_J = 0.125$ or 12.5%.

Fish N has more than one common ancestor. This means that more than one path will determine the inbreeding of fish N. Fish A and fish C are both common ancestors of fish N. A glance at the pedigree reveals that fish A and fish C are the only fish to appear on both sides of fish N's pedigree. Fish J is inbred, but since it does not appear on both sides of fish N's pedigree, its inbreeding value does not contribute to fish N's inbreeding value.

The path that can be traced from fish J to fish M (the parents of fish N) though common ancestor fish C is:

J-I-C-K-M

There are five individuals in this path, so N = 5. Two different paths can be drawn from fish J to fish M (the parents of fish N) though common ancestor fish A:

1. J-H-A-M
2. J-I-A-

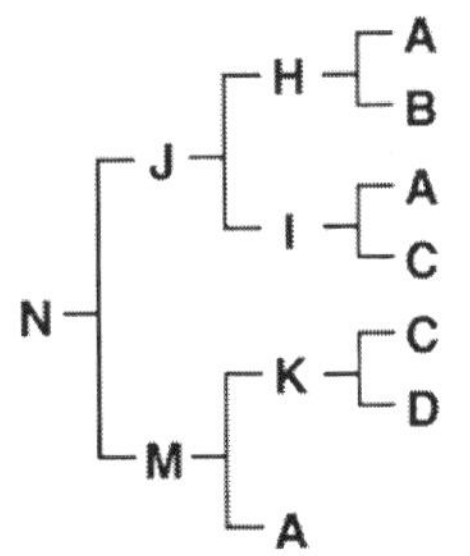

Figure: Pedigree

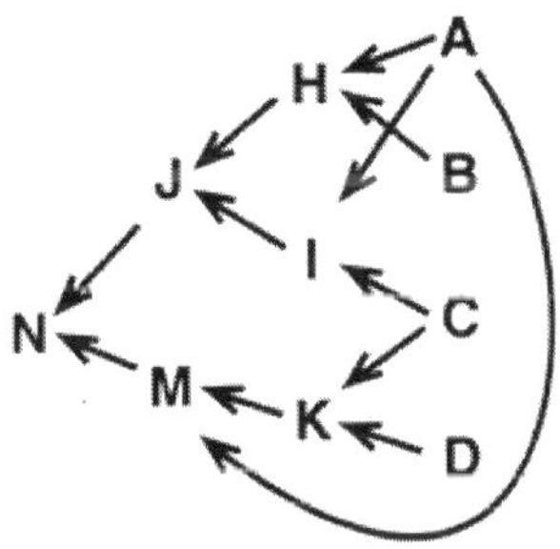

Figure: Path Diagram

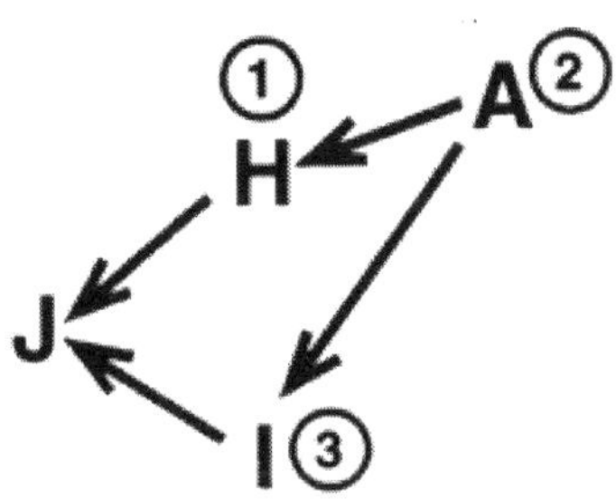

Figure: Path From J To Common Ancestor A

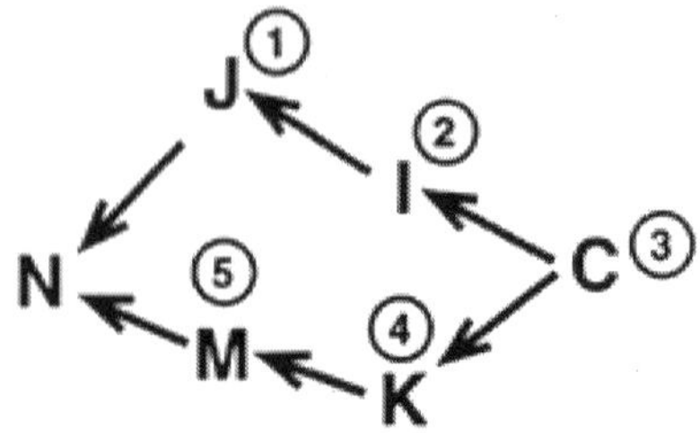

Figure: Path From N To Common Ancestor C

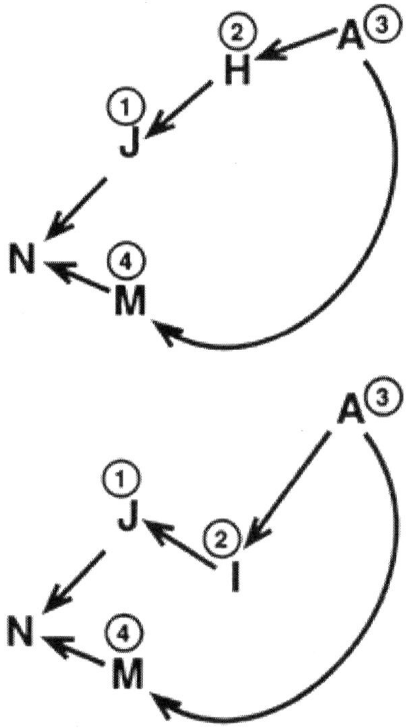

Figure: Paths From N To Common Ancestor A

A pedigree and the path diagram that is used to calculate inbreeding. The paths that are used to calculate inbreeding in fish J and fish N are shown.

There are four individuals in each path, so N = 4 for both. Since neither common ancestor is inbred, the simplified formula can be used. The only difference that occurs when there is more than one common ancestor or when more than one path exits for a particular common ancestor is that the inbreeding value is determined by adding the value derived from each path:

$$F_N = (0.5)^5 + (0.5)^4 + (0.5)^4$$

$$F_N = 0.03125 + 0.0625 + 0.0625$$

$$F_N = 0.15625$$

Thus, $F_N = 0.15625$ or 15.62%.

There are two important rules about tracing paths:

Rule 1: You cannot retrace a path; i.e., you cannot go though any individual twice in a given path.

Rule 2: A path is traced backward from one parent to the common ancestor, and then forward from the common ancestor to the other parent.

The part of the path that goes from one parent to the common ancestor will travel only in a backward direction; i.e., the path it traces will always start at the arrowhead and go towards the shaft. The part of the path that returns from the common ancestor to the other parent will travel only in a forward direction; i.e., the path it traces will always start at an arrow's shaft and go to the arrowhead. A path that travels in both directions, either on the way to the common ancestor or on the way back, is an erroneous path; i.e, two arrowheads or the ends of two shafts cannot touch in a given path. This is why you cannot use the path J-H-A-I-C-K-M when determining the inbreeding of fish N. This path goes forward and backward while going from common ancestor A to parent M; two arrowheads touch when going A-I-C.

Covariance Analysis

In covariance analysis, a pedigree is converted to a covariance table. Each individual in the pedigree is listed at the top of each column and to the left of each row. The parents of each individual are listed at the far left of the table. If one or both parents are unknown, that information is represented by a dash in the parents' column.

The cells that make up a covariance table will contain either the covariance values between two individuals or the covariance value of an individual. The cells that form the diagonal that starts with the upper left cell and ends with the lower right cell and that are the intersections produced by each individual's row and column (e.g., cells AA, BB, CC, etc.) are the cells that contain individual covariance values. The cells that lie below this diagonal are the cells that contain covariance values between two individuals.

Because all individuals are listed both in the rows and in the columns, the cells above the diagonal and those below the diagonal will contain the same information. It is not necessary to record the information twice, so the cells above the diagonal are left blank.

Table 1. Example of a table of covariance values for brood fish at a fish station. Males are listed across the top, and females are listed along the left. This table is a compilation of Cov_{BI} values among all brood fish, so all cells are filled; the information above the diagonal is not the same as that below the diagonal, and the diagonal values are not Cov_{Ind} values. This table can be used to predict the inbreeding that will be produced by any mating combination. The inbreeding that will be produced by any mating is simply half the Cov_{BI} value.

Females	*Males*									
	1	2	3	4	5	6	7	8	9	10
1	0.0	0.125	0.125	0.125	0.0	0.5	0.5	0.5	0.25	0.25
2	0.75	0.75	0.0	0.0	0.25	0.0	0.625	0.625	0.0	0.25
3	0.0	0.0	0.0	0.25	0.0	0.0625	0.0625	0.625	0.0	0.25
4	0.125	0.125	0.5	0.5	0.5	0.0625	0.0625	0.25	0.25	0.25
5	0.0	0.25	0.0	0.0	0.75	0.75	0.75	0.0	0.625	0.0
6	0.0	0.125	0.125	0.25	0.0	0.625	0.625	0.625	0.625	0.0
7	0.25	0.125	0.125	0.0	0.5	0.0	0.0	0.625	0.625	0.625
8	0.0	0.25	0.25	0.0	0.5	0.0	0.0	0.625	0.0625	0.0625
9	0.0	0.25	0.25	0.0	0.5	0.0	0.0	0.625	0.125	0.125
10	0.0	0.025	0.25	0.0	0.0	0.625	0.625	0.0	0.5	0.5

Individual inbreeding values can be determined only if a farmer has detailed breeding records which can be turned into pedigrees. There are many reasons why farmers and hatchery managers should maintain breeding records. The ability to determine individual inbreeding values is but one reason. If such information exists, it is easy to calculate individual inbreeding values and to use them to predict future inbreeding values from various mating combinations.

Individual inbreeding values can be determined by two techniques: path analysis and covariance analysis. Both techniques provide accurate results, so the method that is used is determined by personal choice. Path analysis requires less work, but following paths can be tricky, so it is easier to make a mistake using this method. Covariance analysis is more tedious, but the math is very simple, and the completed table can be used to quickly predict any inbreeding that would be produced by mating any fish that are included in the table.

Both techniques clearly demonstrate an important fact: An individual is inbred if and only if its parents are related through one or more common ancestors. Both parents can be highly inbred, but if they are not related, their offspring will have F = 0.0. Fish J and fish U are inbred; $F_J = 0.25$ and $F_U = 0.375$. However, the mating of fish J and fish U will produce offspring with no inbreeding, because fish J and fish U are not related through a common ancestor. F_Z = 0.0 because no fish appears on both the maternal and paternal sides of fish Z's pedigree.

This fact has important consequences for brood stock management and for breeding programmes. Inbreeding can be reduced to zero in a single generation by mating unrelated fish. Even if inbreeding has reached levels that cause problems, inbreeding can be eliminated and prevented in subsequent generations simply by examining pedigrees and by mating unrelated fish.

This can also be accomplished by acquiring fish from another hatchery and by making hybrids. If the fish from the two hatcheries are unrelated, the F_1 hybrids will have inbreeding of zero. F_1 hybrids produced by this technique always have inbreeding of zero, provided the parents are unrelated. In fact, many hybrid seeds that are produced commercially are produced by crossbreeding highly inbred, but unrelated, lines. This reduces inbreeding to zero in the plants that will be grown and also produces some hybrid vigour, as well as uniform crops.

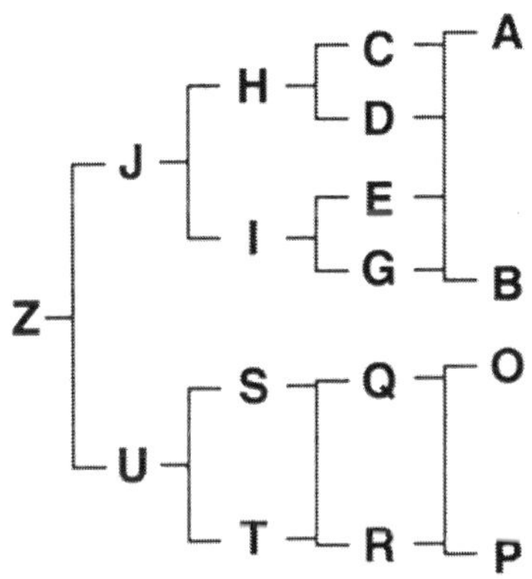

Figure: Pedigree illustrating the fact that the mating of two inbred, but unrelated, fish will produce offspring with no inbreeding: F_J = 25% and F_U = 37.5%, but because they are not related, F_Z = 0%.

Calculating Average Inbreeding Values in Hatchery Populations

Most farmers and hatchery managers cannot calculate individual inbreeding values for fish that they culture, because individual fish cannot be identified and because pedigrees are not recorded. However, this does not mean that inbreeding values cannot be determined for hatchery fish. Even though individual inbreeding values cannot be calculated, population averages can be and should be determined.

Average inbreeding values should be determined for all hatchery populations every generation, and the acquisition and use of these data should be an integral part of hatchery management. The average inbreeding value is as important as average harvest weight or any other data that can be gathered. The average inbreeding value of a population is determined from the population's effective breeding number (N_e). Effective breeding number is one of the most important bits of information that can be gathered about a hatchery population, and N_e should be determined every generation for all hatchery populations. The first part of this chapter will define N_e, will describe what factors influence N_e, and will then show how N_e is calculated. Effective breeding number is traditionally calculated by counting the number of males and number of females that produce viable offspring. This direct approach will be the technique described in this chapter.

Recently, protocols have been developed to assess N_e indirectly by examining changes in gene frequency and by assessing something called "linkage disequilibrium," which is the difference between the actual and expected co-occurrence of two alleles at two loci. In some cases, these indirect procedures may provide a more accurate estimate of N_e, because they account for differential reproduction and offspring survival, which can have a marked impact on N_e. However, these techniques are more expensive because they require detailed biochemical work and they also require highly trained scientists who are skilled at biochemical population genetics, requirements that are not available at most farms or hatcheries.

Consequently, the traditional approach of enumerating the males and females that spawn and that produce viable offspring will be the only technique that will be described in this chapter. This approach is easy and the math is relatively simple. Those who wish to learn about the indirect procedures will find papers that describe these techniques in Recommended Reading.

The techniques outlined in this chapter are fairly simple. Subsistence-level farmers do not need to know how to determine N_e and average inbreeding, but any farmer or hatchery manager who wants to increase yields and productivity or to manage populations that are being cultured in order to restock lakes and rivers should learn how to determine these values and should incorporate these protocols into yearly work plans.

Effective breeding number and the average inbreeding value are important population descriptors, because they help explain trends in yield, fecundity, and other important production phenotypes. These values also enable a farmer to predict if problems will occur as a result of inbreeding depression or the loss of genetic variance. Proper brood stock management cannot be accomplished without this information.

Effective Breeding Number (N_e)

The N_e of a hatchery population is one of the most important bits of information about the population. Unfortunately, most hatchery managers do not know what N_e is, do not know how to determine N_e, and do not know how N_e influences inbreeding and brood stock management. If asked to describe the size of their hatchery population, most farmers or hatchery managers would produce a census or an approximate number. Those with good records would probably be able to give the number of male and female brood fish. This information is used to determine the number of fingerlings that can be produced, to calculate the amount of feed that must be ordered, or to estimate yield. As important as this information is, it does not describe the population genetically. In order to describe a population genetically, one must determine N_e. Effective breeding number is one of the most important concepts in brood stock management, because it gives an indication about the genetic stability or genetic health of the population. This is because N_e is inversely related to inbreeding and to genetic drift. If hatchery populations were infinitely large, an understanding of N_e would be unnecessary. However, hatchery populations are usually small and are often closed. A closed population is one where immigration (the introduction of fish from another population) is not allowed; consequently, fish from other populations are not allowed to mate with or hybridize with fish from a closed population. Hatchery managers often maintain closed populations for various reasons; chief among them is the desire to minimise health problems by preventing the introduction of diseases that often accompany acquired fish.

When working with a closed, finite population, the best way to describe it is not by total number of fish, but by N_e. Effective breeding number is determined by the number of male and female brood fish that produce viable offspring, the sex ratio of the brood fish that spawned, the variance of family size, and the mating system that is used. In most situations where fish cannot be identified and where mating is random (fish are paired without regard to phenotypic value, or fish swim free in a pond and choose their own mates), N_e can be determined by using the following formula:

$$Ne = \frac{4(\textit{number of males})(\textit{number of females})}{\textit{number of} + \textit{number of females}}$$

where number of males and number of females are the number of male and female brood fish that produce viable offspring. If all offspring for a brood fish die, that male or female is not included when determining N_e. If matings and offspring production cannot be monitored, N_e cannot be determined. If fish spawn in ponds and eggs are allowed to hatch in the ponds and if offspring are not harvested until they are mixed schools of fry or fingerlings, N_e will be difficult, if not impossible, to determine. The formula shows that N_e is determined both by the number of male and female brood fish and by the sex ratio. For example, if 53 female brood fish produced eggs and if 25 males were used to fertilize those eggs and if all brood fish produced viable offspring, N_e is:

$$Ne = \frac{4(53)(25)}{53+25}$$

$$Ne = \frac{5300}{78}$$

$$N_e = 67.95$$

This example illustrates a fundamental concept of brood stock management: The genetic size of the population (N_e) and the number of fish that produce offspring are not always the same; the genetic size is usually smaller. Seventy-eight brood fish produced offspring, but N_e was only 67.95. The reason N_e was smaller is because the sex ratio was skewed; in this example, the sex ratio was 2.12 females: 1 male (53 females:25 males).

Effective breeding number and the number of brood fish that produce offspring will be the same only when the sex ratio is 1:1. In the above example, there were 78 total brood fish, so a 1:1 sex ratio would have been 39 females:39 males. If these numbers were used to produce offspring, N_e would have been:

$$Ne = \frac{4(39)(39)}{39+39}$$

$$Ne = \frac{6084}{78}$$

$$N_e = 78$$

The effect of number of males and females and of the sex ratio in determining N_e. The information presented earlier clearly shows that the best way to increase N_e is to increase the number of males and the number of females that spawn and produce viable offspring and to bring the sex ratio closer to a 1:1 ratio. Increasing one sex while keeping the other sex at a fixed number has little effect on increasing N_e after a certain point; this can be seen by examining the curves for 1, 2, 10, and 25 males.

Unequal reproductive success can affect N_e. If some fish produce thousands of viable offspring while others produce only dozens, N_e will be less than expected. Variance in family size is a major reason why N_e is often smaller than expected. Variance in family size effects N_e as follows: When there are single pair matings, variance in family size is equal for both sexes, and N_e becomes:

$$NeUR = \frac{4(Ne)}{\text{var}\,iance\,of\ offspring\ production + 2}$$

Where N_{eUR} is effective breeding number when there is unequal reproductive success and variance in offspring production is variance in family size.

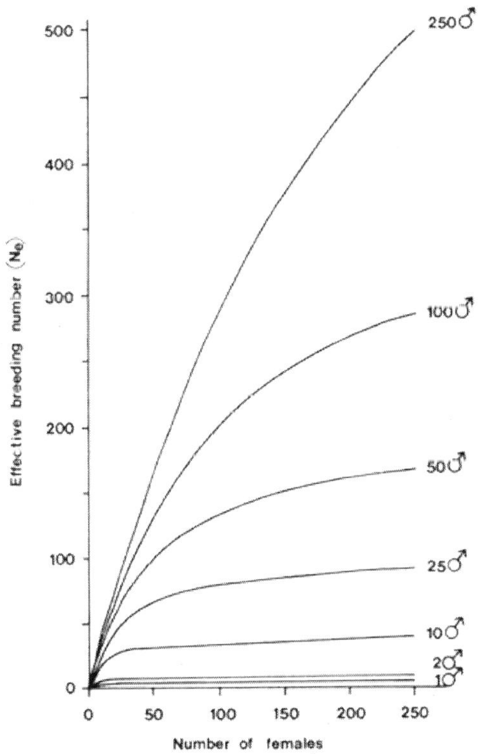

Figure: . Effective breeding numbers produced by various combinations of males and females.

From: Tave, D. 1993. Genetics for Fish Hatchery Managers, 2nd ed. Van Nostrand Reinhold, New York, New York, USA.

For example, if N_e in a population is 100, there are single pair matings (50 males and 50 females), and the variance in family size (variance in offspring production) is 5, N_e becomes:

$$NuUR = \frac{4(100)}{5+2}$$

$$NuUR = \frac{400}{7}$$

N_{eUR} = 57.14

The variance in family size reduced predicted N_e by 43%, and yielded an effective breeding population of just 28 males and 28 females.

When males mate with more than one female and/or females mate with more than one male, there is unequal reproductive success for the two sexes, and N_e becomes:

$$NuUR = \frac{8(Ne)}{\text{var}\,iance\ of\ female\ offsprig\ production + \text{var}\,iance\ of\ male\ offspring\ prodi}$$

Family size often assumes what statisticians call a "Poisson distribution," and when this occurs the mean and the variance are the same. If this occurs, mean family size can be used.

The N_e that has been calculated in the previous examples is that for a single generation. It is important to determine N_e for every generation, because the values are not independent. The average N_e for a series of generations is determined by calculating the harmonic mean:

$$\frac{1}{Nemean} = \frac{1}{t}\left(\frac{1}{Ne1} + \frac{1}{Ne2} + .. + \frac{1}{Net}\right)$$

where: N_e mean is the mean effective breeding number over t generations; and N_{e1}, N_{e2}, and N_{et} are the effective breeding numbers in generations 1, 2, and t, respectively. The math needed to determine mean N_e is fairly simple and can be done "by hand"; however, inexpensive hand-held calculators are preprogrammed to determine inverses (the "1/x" key) so their use reduces this formula to simple arithmetic.

Because the mean is determined by the harmonic average, the generation with the smallest N_e will have a controlling influence on average N_e. This means that if a disease or other calamity reduces

N_e, subsequent increases in N_e will have little effect on increasing the average.

For example, if the N_e's for five generations are 50, 10, 65, 85, and 100, the mean N_e is:

$$\frac{1}{Ne\,mean} = \frac{1}{5}\left(\frac{1}{50} + \frac{1}{10} + \frac{1}{65} + \frac{1}{85} + \frac{1}{100}\right)$$

$$\frac{1}{Ne\,mean} = 0.2(0.02 + 0.1 + 0.01538 + 0.01176 + 0.01)$$

$$\frac{1}{Ne\,mean} = 0.031428$$

$N_{e\ mean}$ = 31.82

In this example, N_e went below 50 only once, but mean N_e was 31.82 because N_e dropped to 10 for a single generation. Even though N_e was 85 and 100 in the final two generations, if just 16 males and 16 females had been spawned each of the five generations, mean N_e would have been slightly larger ($N_{e\ mean}$ = 32).

This example illustrates one of the most important concepts of brood stock management: N_e must be managed each and every generation. The population's N_e must be considered to be one of its most important descriptors. If N_e declines to a low level for just a single generation, this will have a devastating impact on mean N_e, even if N_e is maintained at high levels before and after the single-generation reduction. When a harmonic mean is used to determine the mean, the generation with the lowest value has a disproportionate effect on the mean.

Inbreeding

Once N_e has been determined, a simple formula can be used to calculate the average inbreeding value in the population:

$$F = \frac{1}{2Ne}$$

where F is the average inbreeding value in the population. For example, if N_e = 100, the average inbreeding value for the fish in the population is:

$$F = \frac{1}{2(100)}$$

F=0.005

This value is assigned to every offspring produced by those brood fish.

The formula shows that F is inversely related to N_e: a large N_e produces a small F; a small N_e produces a large F. The relationship between F and N_e in this formula clearly demonstrates why it is important to calculate N_e and why management of N_e is of paramount importance in the management of hatchery populations.

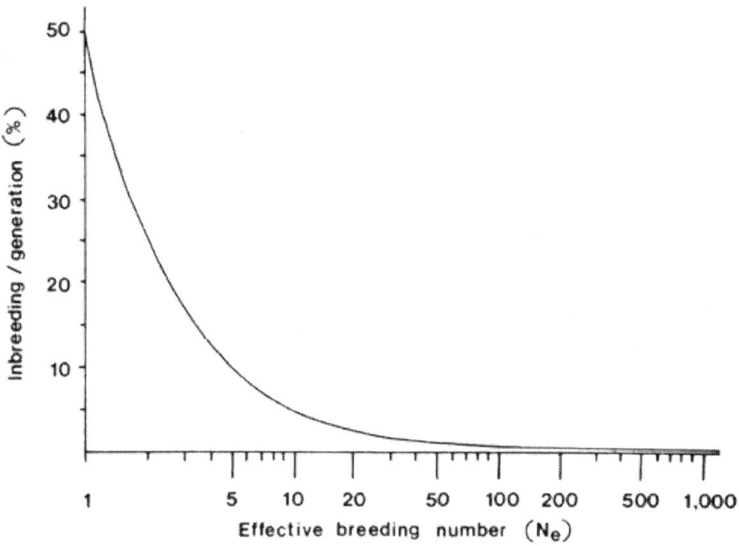

Figure: The relationship between effective breeding number (N_e) and average inbreeding (F) in a population. This is the average inbreeding produced by a single generation of mating.

From: Tave, D. 1990. Effective breeding number and broodstock management: I. How to minimise inbreeding. Pages 27–38 in R.O. Smitherman and D. Tave, eds. Proceedings Auburn Symposium on Fisheries and Aquaculture. Alabama Agricultural Experiment Station, Auburn University, Alabama, USA.

The inbreeding that is calculated in this manner is genetically identical to that which was determined; it is a measure of the increase in homozygosity that occurred as a result of the mating of relatives. The only difference is: when individual inbreeding values are determined, the values are customized for each individual, based on its family pedigree; when N_e is used to determine inbreeding, the value that is calculated is the population average. Consequently, some fish will be more inbred than the average while others will be less inbred, and there is no easy and inexpensive way to determine which fish are above the mean and which are below. But if pedigrees do not exist, this is the only way to produce an estimate of the inbreeding that has accumulated in the population.

The reason why F and N_e are inversely related is that chance encounters between relatives increase in small populations, and inbreeding is the mating of relatives. When mating is random, closed populations produce consanguineous matings by chance, and the probability of consanguineous matings is inversely related to N_e. For example, if a fish has 50 relatives in a population of 1,000,000, the odds of that fish mating with a relative (assuming there is a 50:50 sex ratio) is 25/500,000 or 0.00005 if mates are chosen at random without regard to pedigree. If a fish has 50 relatives in a population of 1,000 fish, the odds of mating with a relative jumps to 25/500 or 0.05, which is 1,000 times greater. Consequently, the probability of mating with a relative and of producing inbred offspring increases as N_e decreases.

Once inbreeding is produced in a small closed population as a result of a small N_e, the inbreeding itself lowers future N_e in a positive feedback cycle:

$$NeF = \frac{Ne}{1+F}$$

where: N_{eF} is the effective breeding number in an inbred population, and F is the average inbreeding in the population.

The positive feedback illustrated in this formula means that reductions in N_e increase inbreeding, which in turn lowers N_e, which in turn increases inbreeding, etc. Because it is a positive feedback, the system takes a bad situation and continually makes it worse.

Farmers and hatchery managers should determine the N_e of a hatchery population every generation. A population's N_e is one of the most important pieces of information about that population because it describes its genetic size. An understanding of N_e and what it means is of major importance because N_e helps determine the genetic stability of the population in that it determines the average inbreeding in the population and influences genetic drift. Once N_e has been quantified, it can be used to determine the average inbreeding value in the population.

The average inbreeding value in the population is inversely related to N_o, which means that inbreeding can become a problem in small, closed populations, which are typical of those found at most farms and fish culture stations. Once inbreeding occurs, the problem can quickly become worse as inbreeding and N_e are linked in a positive feedback loop. Consequently, small N_e's will produce levels of inbreeding which will cause growth rate, fecundity, and other traits to decline over succeeding generations. It is important to understand what N_e is, how it is determined, and how it affects inbreeding so that it can be managed to prevent unwanted inbreeding from decreasing productivity and profits.

Genetic Drift

Genetic drift is the second important genetic concept that is a function of N_e. Genetic drift is random changes in gene frequency; it is a major factor in evolution and population biology, but most aquaculturists have not heard of it. However, like inbreeding, it too must be considered if hatchery populations are to be properly managed.

Genetic drift can be as important as selection in altering a population's gene pool. The effects of genetic drift can be devastating. Genetic drift can irreversibly alter gene frequencies and eliminate alleles, which can decrease a population's ability to survive or to adapt to an altered environment, and it can preclude future selection. The effect that genetic drift can have on a population's gene pool can make many management goals impossible to achieve.

Effective Breeding Number and Genetic Drift

Genetic drift is random changes in gene frequency that occur because of sampling error. Sampling error can be natural, or it can be manmade. Natural sampling errors are those which occur when earthquakes, floods, landslides, or other natural disasters subdivide a population and isolate small groups of organisms. This process is a major force in the evolution of new species. Manmade sampling errors are inaccurate collections: sampling only a portion of the population, sampling only a single age class, sampling only fish that possess a certain phenotype or that spawn on a particular day, etc.

When a population is sampled, there is a chance that the sample does not accurately reflect the make-up of the population. This inaccuracy can include length, sex ratio, body colour, and gene frequencies. The smaller the sample, the greater the likelihood that inaccuracies in the sample will occur. Changes in gene frequency that occur as a result of sampling error are called "genetic drift."

When culturing fish, the important changes that can occur in gene frequency as a result of genetic drift occur during the creation of the next generation (during spawning season) or during the acquisition of the population. This is when the genes are transferred from parents to offspring (the transfer of genes across time; from one generation to the next) or when they are transferred from one hatchery to another (the transfer of genes across space). As was the case with inbreeding, N_e is the factor that determines the magnitude of genetic drift. The relationship between N_e and genetic drift is:

$$\sigma^2_{\Delta q} = \frac{Pq}{2Ne}$$

Where: $\sigma^2_{\Delta q}$ is the variance in the change of gene frequency, and p and q are the frequencies of alleles p and q for a given gene. The variance of the change in gene frequency is the way genetic drift is measured.

Like average inbreeding value, genetic drift is also inversely related to N_e. A comparison of the two formulae shows that they are similar. This means that large N_e's produce small changes in gene frequency, while small N_e's produce large changes in gene frequency.

Many factors can cause changes in gene frequency; among them are: selection by the farmer or hatchery manager; domestication (selection caused by the hatchery and the techniques used to culture the fish); genetic drift. The changes in gene frequency caused by selection and domestication usually produce genetic improvements, in that the fish grow faster, are more disease resistant, are calmer, accept pelleted feed more readily, have more efficient feed conversions, and are easier to spawn. In contrast, the changes in gene frequency that are caused by genetic drift are random, which means they can be counterproductive.

If the frequency of an allele changes from, say, 0.5 to 0.45 or from 0.4 to 0.3 as a result of genetic drift, the genetic effects on the population might not be that great. The major damage that is caused by genetic drift occurs when the frequency of an allele goes to 0.0; i.e., the allele is lost and no longer exists in the population. The odds of losing alleles via genetic drift are related to their frequencies; i.e., rare alleles (the frequency is low; usually ≤ 0.01) are lost more easily than common ones.

The probability of losing an allele by genetic drift is determined by using the following formula:

$$P = (1.0 - q)^{2N_e}$$

where: P is the probability of losing an allele, and q is the frequency of the allele.

For example, if N_e is 50, the probability of losing an allele whose frequency is 0.1 (q = 0.1) is:

$P = (1.0 - 0.1)^{2(50)}$

$P = (0.9)^{100}$

$P = 0.0000265$

If N_e is 10, the probability of losing the same allele is:

$P = (1.0 - 0.1)^{2(10)}$

$P = (0.9)^{20}$

$P = 0.12158$

These examples clearly show that the probability of losing an allele is inversely related to N_e; the probability of losing the allele was 4,588 times greater when N_e decreased from 50 to 10.

The genetic stability of a closed hatchery population depends on the population's N_e. As was the case with average inbreeding, genetic drift is inversely related to N_e. Consequently, small hatchery populations can cause random changes in gene frequency. The ultimate effect of a small N_e is the loss of alleles via genetic drift. Rare alleles will be lost more easily, but common alleles can also be lost. The loss of genetic variance can produce irreversible damage to a population's gene pool. This loss can prevent future improvements via selection, and it can reduce viability in populations that are stocked in lakes and rivers.

Using Inbreeding to Improve Growth and Other Phenotypes

A farmer or hatchery manager who wants to manage the genetic aspects of his population must know how inbreeding can be used to improve productivity and profits. Inbreeding is one of the three major traditional breeding programmes that breeders have used for centuries to improve animals and plants. While it is not as important as selection or crossbreeding, inbreeding is used to produce genetically improved livestock, plants, and laboratory animals. Inbreeding might be the most important breeding technique used in the production of laboratory animals, because genetically uniform lines of rats, mice, etc. are desired for biological and medical research.

Inbreeding will never be as important in animal husbandry as it is in agronomy, because many plants can be self-fertilized; it is easy to create matings by artificial pollination; and many individuals can be raised cheaply in a small plot. However, inbreeding has been used to create better, faster growing livestock, and new technologies have improved our ability to create and use highly inbred fish to improve a population or for research purposes. Inbreeding programmes should be far easier with fish than with livestock, because many species can be stripped, which enables fish culturists to create mating combinations livestock farmers can only imagine. Additionally, fish are highly fecund when compared to livestock, so different types of inbreeding programmes can be used.

In order to use an inbreeding programme to improve productivity and profits, a farmer or hatchery manager must understand how inbreeding works and what it does. The genetics of inbreeding, and the techniques that are used to calculate individual inbreeding values were outlined. This chapter will explain how inbreeding can be used to improve populations by itself, or how inbreeding can be combined with selection and crossbreeding to improve growth rate and other phenotypes.

Inbreeding Programmes

Although selection and crossbreeding are the breeding programmes that are usually considered when plans are made to improve a population genetically, inbreeding is a third option that can be used to produce good results. Inbreeding is generally shunned because it is a two-edged sword that can mortally wound a population. However, when used properly, inbreeding can be an effective and efficient breeding programme. In general, inbreeding programmes are used when you have superior animals. If you inbreed average animals, you produce average animals. But if you inbreed superior animals, you can create outstanding animals.

Creation of New Breeds

Inbreeding is often used when a new breed, strain, or variety is founded. In many cases it is inevitable and unavoidable. New breeds can be formed as the result of a fortuitous hybrid, or they can originate from a single individual with an unusual or desirable phenotype, as was the case with the Morgan horse, an American breed founded by a stallion named Justin Morgan. When breeds are new and small, inbreeding is inevitable. When breeds or strains are first created, there may be only a few males in the breed, and one is considered to be far superior to all others. When this happens, that male is bred to many females and to a good percentage of his daughters and granddaughters in order to produce a population of animals that resembles him. This is how a breed "type" (a particular body conformation) is created. A second round of inbreeding can occur if only one or two of the male's sons are used. When the size of the breed increases, inbreeding will be less important, but it is the breeding technique that is often used to develop a new breed or strain that breeds true for type.

Linebreeding

Linebreeding is an inbreeding programme whereby an individual is mated to its descendants. Traditionally, a single male is bred to many females. Most linebreeding programmes that are used in livestock husbandry only linebreed males, because fecundity is low and gestation periods are long. However, in aquaculture the fecundity of fish should enable farmers and hatchery managers to use females as well as males in linebreeding programmes.

Linebreeding can be used when the breed is in its infancy to develop the breed, or it can also be used in an established breed or strain when an outstanding animal is discovered. Linebreeding is used to increase an outstanding individual's contribution to a population, especially when a farmer thinks that the animal is so superior that it is unlikely that he will ever find a better one.

Linebreeding is used by many livestock farmers, even when they say that they do not want inbred animals. Linebreeding is often not considered to be inbreeding, because the major goal is not to produce inbred animals but to increase an outstanding individual's contribution to the population and to succeeding generations. However, this will occur only because linebreeding is inbreeding.

Even if inbreeding reaches fairly high levels in a population, if the inbreeding is planned it can be used to produce outstanding individuals, and linebreeding is an inbreeding programme that is designed to accomplish that goal. Inbreeding depression is a measure of the population, not of an individual. Inbreeding depression is the difference between the mean of the inbred group and the mean of a control population with no inbreeding. Since inbreeding depression is a populational value, it represents the average value, which means an individual with a high level of inbreeding can be outstanding. If a breeder linebreeds a superior male only with superior females, this programme can quickly produce a superior population.

Linebreeding is the breeding programme that is needed if a farmer wants to keep the relationship of the next generation with a particular ancestor at a high level. The genetic contribution that a fish makes to a population is halved each generation: 50% in its offspring, 25% in its grandchildren, 12.5% in its great-grandchildren, etc. After seven generations, a fish contributes less than 1% of its descendants' genes. Linebreeding is the breeding programme that can be used to reverse this trend. For example, if a fish is mated with its grandchildren (the offspring would be both its great-grandchildren and its children), its contribution to those offspring would rise would rise from the normal value of 12.5% to 62.5%.

If the male that is being linebred mates only once or twice with his descendants, the linebreeding is considered to be mild linebreeding. If the male that is being linebred is repeatedly mated to his female descendants or is mated to them using a regular pattern, the linebreeding is considered to be intense linebreeding.

Intense linebreeding is a pre-biotechnology breeding programme that can be used to create a "near clone." By repeatedly mating a male to his female descendants a farmer can create an "almost identical" genetic twin of the male.

Use of a Herd Bull

This type of inbreeding programme is a variation on linebreeding. In this type of inbreeding programme, a farmer usually purchases an outstanding bull and uses it as a "herd bull"; i.e., it is the only male in the herd. This bull is mated to all females and many female descendents. This inbreeding programme is used to bring a herd of animals up to the herd bull's phenotypic standards.

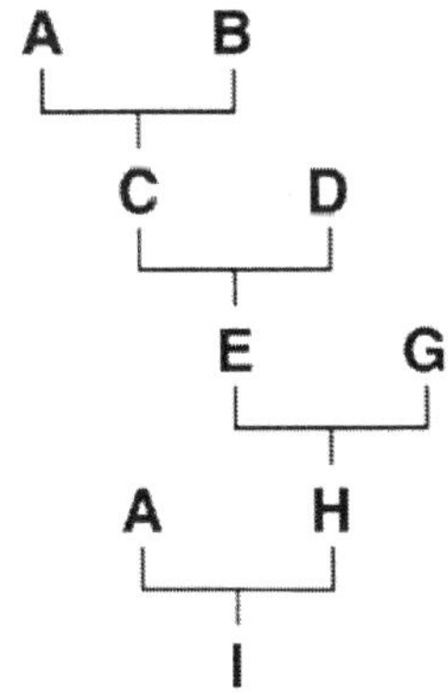

Figure: An example of a linebreeding programme. This example can be considered to be mild linebreeding. Individual A is brought back and mated with his great-grandchild. Individual A can be a male or female, but linebreeding usually uses males. Consequently, individual A is the father and great-great-grandfather of individual I. This is used to increase the percentage of A's genes in his descendants; 56.25% of I's genes come from A; F_I = 6.25%.

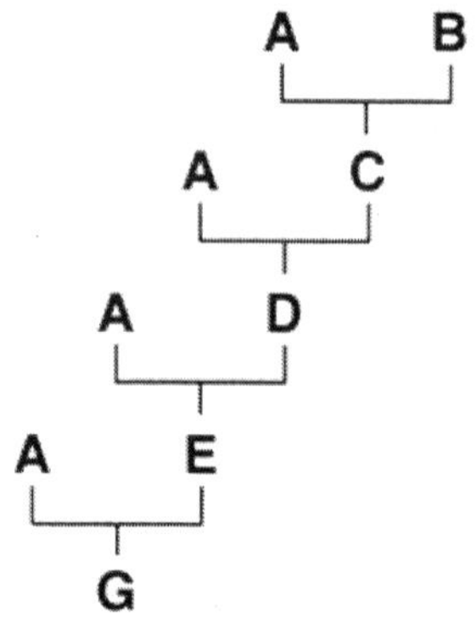

Figure: An example of a linebreeding programme. This can be considered to be intense linebreeding. Individual A is repeatedly mated to his daughters. In this example, individual A is the father, grandfather, great-grandfather, and great-great-grandfather of individual G. This breeding programme is used to increase the percentage of A's genes in his descendants; 93.75% of G's genes come from A; F_G = 43.75%.

This breeding programme will create herds that are highly inbred, so some farmers use several herd bulls to reduce the level of inbreeding and thus minimise inbreeding depression. The inbreeding that will be produced by this inbreeding programme can be determined by the following equation:

$$F = \frac{1}{8(number\ of\ females)} + \frac{1}{8(number\ of\ males)}$$

where: number of males and number of females are the number that produce viable offspring. The formula shows that inbreeding is controlled by the sex that is least numerous, so the rate of inbreeding is governed by the number of herd bulls. For example, if 50 females are used, the inbreeding that will be produced per generation when 1, 2, 3, 4, and 5 herd bulls are used is:

Number of Herd bulls mated to 50 Females	*Inbreeding per Generation*
1	12.75%
2	6.50%
3	4.42%
4	3.38%
5	2.75%

Inbreeding to Expose and Cull Detrimental Recessive Alleles

Inbreeding can be used as an extreme form of progeny testing to expose detrimental recessive alleles and eliminate families that carry these undesired genetic bombs. Progeny testing is a selective breeding programme that is often used to accomplish this, but progeny testing is usually used to eliminate only one recessive allele. Progeny testing is a breeding programme where fish with the dominant phenotype are mated to a test fish (one that is homozygous recessive) to identify and cull the heterozygotes and to identify and save the homozygotes.

By mating relatives and developing highly inbred families, a farmer can unmask many detrimental recessive alleles simultaneously and cull families that carry many hidden genetic defects. Since most animals carry several to several dozen detrimental recessive alleles that are hidden in the heterozygous state, this type of breeding programme is a severe form of progeny testing, but families that exhibit no problems are genetically superior and defect-free. Because this type of breeding programme is so severe, many inbred families must be created in order to identify and save the few that will be defect-free. Although this is costly, it is a breeding programme that will produce outstanding animals, in terms of qualitative phenotypes, within a few generations.

Inbreeding to Improve the Results of between-Family Selection

Inbreeding can be used to improve the results of selection when the h^2 for a trait is small. When h^2 is ≤ 0.15, it is difficult to improve the population by individual selection, because most of the phenotypic differences that can be measured are not heritable. When this occurs, between-family selection is more efficient because this selective breeding programme magnifies the heritable differences (those due to V_A) by minimising environmental effects (V_E).

When h^2 is small, inbreeding can be used to create inbred families which will further magnify the heritable differences among the families. Inbreeding makes it easier to assess the heritable differences among families, by minimising some of the non-heritable sources (V_D, V_I, and V_E). Inbreeding does not change the absolute amount of V_A; it changes relative amount, which improves the breeder's ability to identify and save families which are superior because of V_A. This makes between-family selection more efficient.

Creation of Inbred Lines for Crossbreeding Programmes

Once a breed is established, perhaps the most important use of inbreeding is to develop inbred lines that will be used in crossbreeding programmes to produce outstanding hybrids for grow-out. Crossbreeding and inbreeding are mating extremes along a continuum: inbreeding is the mating of animals that are more closely related than the average in a population, and crossbreeding is the mating of animals that are less related than the average in a population. The genetics that controls phenotypic expression for both inbreeding and crossbreeding programmes is a function of heterozygosity, which means it is a function of V_D. Inbreeding can be used to develop inbred lines that produce better hybrids-fish that exhibit more hybrid vigour. The creation of inbred lines improves the results of hybridization by "stretching" the genetic distance between lines; i.e., inbreeding can be used to create opposite types of homozygosity, and this will enable hybrids to be more heterozygous and hopefully to exhibit more hybrid vigour. Additionally, the development of inbred lines can be used to fix one or two qualitative traits in each line, so all F_1 hybrids will have two or more desired qualitative traits.

There are three ways this type of breeding programme can be conducted: The first two types hybridize inbred lines. In one, inbred lines are created in two strains that were previously shown to produce outstanding F_1 hybrids. The other option is to create inbred lines in a number of strains and then hybridize them in an attempt to discover a combination that works. The more efficient way is to create inbred lines in two strains that were previously shown to produce outstanding F_1 hybrids. This approach makes sense, because producing superior F_1 hybrids is a hit-or-miss proposition. You cannot predict which mating combinations will nick and produce great hybrids and which mating combinations will produce duds. If you create inbred lines in two strains that have already been shown to produce outstanding F_1 hybrids, it is likely that some of the inbred line combinations will also produce better F_1 hybrids.

Inbreeding can also be used to create an inbred line that will be used in a topcrossing programme. Topcrossing is a crossbreeding programme in which an inbred line is hybridized with a randomly bred line to produce F_1 hybrids. This type of breeding programme has been traditionally used by animal breeders to produce outstanding animals for judging contests. Even though inbreeding can be used to improve a population, it does cause inbreeding depression. Fecundity is often adversely affected by inbreeding. When fecundity declines, it may not be possible to produce enough F_1 hybrids for grow-out. This problem can be circumvented by developing four inbred lines and mating them to create di-hybrids.

The first step in this breeding programme is the creation of four inbred lines. The second step is the creation of two F_1 hybrids. The third step is the mating of the F_1 hybrids to produce a di-hybrid. The creation of the F_1 hybrids reduces F to 0%, so fecundity will return to normal. The F_1 hybrids are used to produce the seeds or animals that farmers will use for grow-out. This is how modern agribusinesses produce seed corn. If each inbred line is used to fix one or two qualitative phenotypes, the di-hybrids will exhibit heterosis, and all will also have several desired qualitative phenotypes.

Inbreeding Produced When Relatives Mate

The ability to use inbreeding in some of these breeding programmes depends on a farmer's or hatchery manager's ability to produce inbred lines quickly and efficiently. This can be accomplished only if breeders know how much inbreeding will be produced by a particular mating programme and the rate at which the inbreeding is created. The amount of inbreeding that will be produced by a given inbreeding programme depends on the relationship that exists between the fish that are mated. Consequently, it is important to know how much inbreeding is produced when relatives mate and produce offspring. This information can then be used to develop regular systems of inbreeding that will enable a farmer to produce predictable levels of inbreeding in a predictable time frame. This section will show how much inbreeding can be produced when relatives are allowed to mate. In order to do this, it will be assumed that no previous inbreeding exists, so the amounts that are mentioned are those that are produced when inbreeding first occurs. The values presented, show that the matings of close relatives, such as parents with offspring and brothers with sisters, produce fish with high levels of inbreeding (25%) and that the inbreeding produced steadily decreases as the distance between relatives increases. For example, matings between third cousins, matings between fourth cousins, and matings between fifth cousins produce very little inbreeding (0.39%, 0.097% and 0.024%, respectively). Fifth cousins share so few genes that the inbreeding produced when they mate is insignificant; thus, on a practical level, many geneticists consider fifth cousins to be unrelated.

The inbreeding values listed, explain why first cousin marriages are forbidden in many cultures while second cousin marriages are sanctioned. First cousin marriages produce children with F = 6.25%, while second cousin marriages produce children with F = 1.56%.

One type of cousin marriage will produce a large amount of inbreeding. A double first cousin marriage (the offspring from the marriages of a pair of sibs that married a pair of sibs; first cousins are individuals that are not sibs but that share two grandparents in common; double first cousins are individuals that are not sibs but that share all four grandparents in common-individuals 22 and 23). Double first cousin matings produce as much inbreeding as half-sib, grandparent-grandchild, and aunt-nephew (or uncle-niece) matings-12.5%. It should be obvious that the mating of unrelated individuals such as 13 and 14 would produce offspring with no inbreeding, but other matings which would usually be considered to be between relatives would also produce offspring with no inbreeding. For example, if individuals 9 and 27 were mated, the offspring would have F = 0%, even though this would be considered to be an uncle-niece marriage. The reason why the 10 × 26 mating produces offspring with F = 12.5% while the 9 × 27 mating produces offspring with F = 0% is individual 9 is individual 27's uncle by marriage. Individuals 9 and 27 share no genes in common; they have no common ancestor and are not related, which is the requirement for the production of inbred offspring. The 9 × 27 mating might be incestuous in most societies, but it produces no inbreeding.

Regular Systems of Inbreeding

The relationships illustrated in Figure and the amount of inbreeding produced by matings among those individuals can be used to develop regular systems of inbreeding in order to produce inbred lines of fish for breeding purposes. The reason why regular systems of inbreeding need to be used is simple: breeding programmes are expensive and take years, so random, haphazard breeding systems that produce unknown and unpredictable amounts of inbreeding are expensive lessons in poor planning and futility. All breeding programmes require meticulous planning, and inbreeding is no exception. If farmers or hatchery managers are going to use inbreeding to improve a population, they must know how to conduct regular systems of inbreeding. Regular systems of inbreeding enable a farmer or hatchery manager to know what matings must be made generation after generation, and this allows them to predict the amount of inbreeding that will be produced each generation.

There are a number of regular systems of inbreeding. Some are relatively simple, but others are so complex or produce so little inbreeding that their use is impractical. A major goal of an inbreeding programme is to produce F = 40–50% in a number of families and to do it fairly rapidly. The only liability of inbreeding programmes that quickly produce large levels of inbreeding is that inbreeding depression will reduce viability and fecundity, and this will cause some of the inbred lines to go extinct. Consequently, many inbred lines must created so that some will survive the ravages of inbreeding depression. There are six types of matings that produce large amounts of inbreeding. Some of these matings are difficult to make generation after generation (e.g., grandparent-grandchild), which limits their use in practical inbreeding programmes. Relatively simple regular systems of inbreeding that will produce F = 40–50% in four to five generations are: parent-offspring, full-sib, half-sib, and double first cousin.

Regular systems of inbreeding divide a population into a series of inbred families. This occurs because the only way a regular system of inbreeding can be maintained is by continually mating a specific combination of relatives, which means mating fish within a family. The size of the family is determined by the inbreeding programme: the closer the relationship between mates, the smaller family size can be.

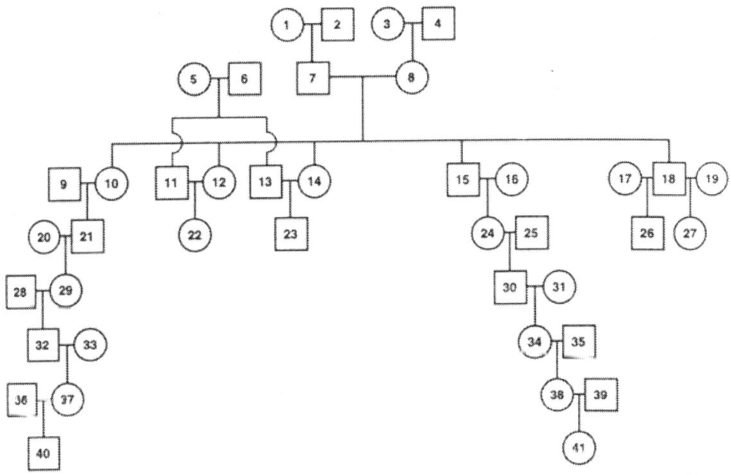

Mating	*Individuals Infamily Tree*	*Produced By Mating*
Parent-Offspring	7 × 10	25.0%
Brother-Sister (full sibs)	10 × 15	25.0%
Half-brother-Half-sister (half sibs)	26 × 27	12.5%
Grandparent-Grandchild	2 × 10	12.5%
Aunt-Nephew (or Uncle-Niece)	10 × 26	12.5%
Double first cousins	22 × 23	12.5%
Great-grandparent-Grandchild	21 × 37	6.25%
First cousins	21 × 24	6.25%
Second cousins	29 × 30	1.56%

Third cousins	32 × 34	0.39%
Fourth cousins	37 × 39	0.097%
Fifth cousins	40 × 41	0.024%
Unrelated	13 × 14	0.0%

Figure: A family tree with 41 individuals. Males are represented by squares and females by circles. This family tree is used to illustrate the amount of inbreeding that would be produced by matings between relatives. Possible mating combinations and the inbreeding that would be produced by these matings range from F = 0% to F = 25%. Individuals 1, 2, 3, and 4 have F = 0% and are not related, and they are not related to individuals 5 and 6. Individuals 5 and 6 have F = 0% and are not related.

Parent-Offspring Inbreeding Programme

A breeding programme of parent-offspring matings is an inbreeding programme that is traditionally used to quickly produce highly inbred animals. A single generation of parent-offspring mating will produce fish with F = 25%, and three generations will produce fish with F = 50%; eventually, F will approach 100%. A parent-offspring inbreeding programme is illustrated. Five generations of parent-offspring matings, along with the inbreeding that is produced, is illustrated. The inbreeding that is created by a parent-offspring inbreeding programme is contrasted to that which can be created by other regular systems of inbreeding.

Table: Percent inbreeding produced by regular systems of inbreeding: parent-offspring; full-sib; half-sib; double first cousin. There are two half-sib inbreeding programmes: a single male is mated to two half-sisters each generation (A); a single male is mated to many half-sisters each generation (B). Generation 1 is the first generation in which inbreeding is produced.

Generation	*Parent-offspring*	*Full-sib*	*Half-sib (A)*	*Half-sib (B)*	*Double first cousin*
0	0.00	0.00	0.00	0.00	0.00
1	25.00	25.00	12.50	12.50	12.50
2	37.50	37.50	25.00	21.88	18.75
3	50.00	50.00	34.38	30.47	25.00
4	59.38	59.38	42.97	38.09	31.25
5	67.19	67.19	50.39	44.87	36.72
6	73.44	73.44	56.84	50.92	41.80
7	78.52	78.52	62.45	56.30	46.48
8	82.62	82.62	67.23	61.09	50.78
9	85.94	85.94	71.58	65.35	54.74
10	88.62	88.62	75.28	69.15	58.37
∞	100.00	100.00	100.00	100.00	100.00

This type of inbreeding programme is simple and requires the fewest numbers of individuals in each inbred family. Only two individuals are needed in each generation-a parent (usually the younger parent) and an offspring. Consequently, this is the least expensive regular inbreeding programme. The major liability of this inbreeding programme is that it is often difficult to mate parents with their offspring generation after generation. A number of fish species spawn only once and then die. When this happens, the only way parent-offspring matings can occur is by using cryopreserved gametes.

Full-Sib Inbreeding Programme

Full-sibs are brothers and sisters that share two parents. This regular system of inbreeding is as effective in producing inbreeding as parent-offspring matings, and it is equally simple and inexpensive. A breeding programme of full-sib matings is illustrated; five generations of full-sib matings along with the inbreeding that is produced is illustrated. As was the case with parent-offspring matings, full-sib matings produce F = 25% in the first generation and F = 50% after three generations; eventually, F will approach 100%.

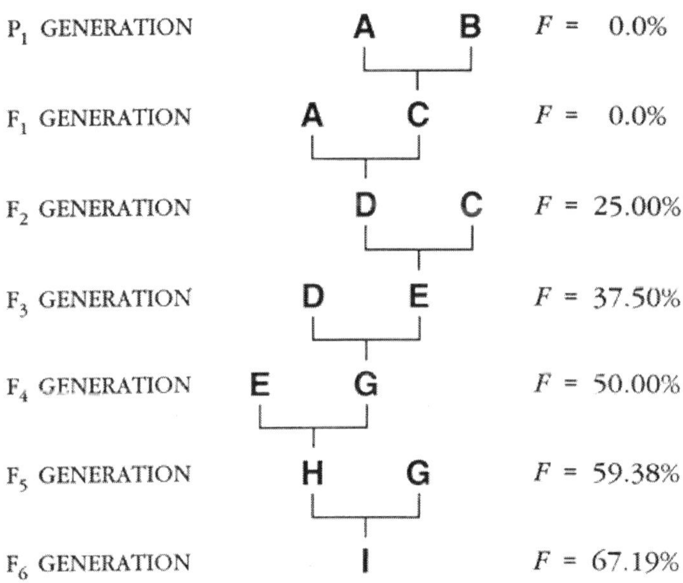

Figure: A regular inbreeding programme of parent-offspring matings. Five generations of parent-offspring matings and the inbreeding produced by this breeding programme are illustrated. The first generation of parent-offspring mating occurs in the F_1 generation, and their offspring (F_2 generation) is the first generation of inbred fish. Fish in the P_1 generation have F = 0% and are not related. In subsequent generations, inbreeding values apply only to individuals which are produced in that generation; e.g., F_c = 0%, not 25%.

Only two individuals are needed per inbred family per generation, but more can be used. Even though family size of parent-offspring and full-sib inbreeding programmes can be identical, a full-sib inbreeding programme is usually less expensive, because when parents are mated to offspring fish have to be cultured for two spawning seasons. When sibs are mated, broodstock can be turned over more rapidly. A major difference between regular parent-offspring and full-sib inbreeding programmes is the fact that it is far easier to make full-sib matings generation after generation. Because full-sib matings are rather easy to make and because this type of inbreeding programme quickly produces large levels of inbreeding, full-sib matings is probably the most commonly used regular system of inbreeding.

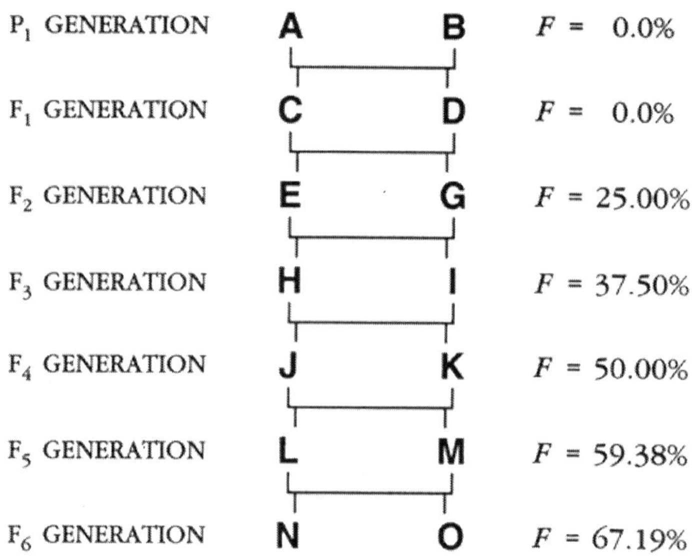

Figure: A regular inbreeding programme of full-sib (brother-sister) matings. Five generations of full-sib matings and the inbreeding produced by this breeding programme are illustrated. The first generation of full-sib mating occurs in the F_1 generation, and their offspring (F_2 generation) is the first generation of inbred fish. Fish in the P_1 generation have F = 0% and are not related.

Half-Sib Inbreeding Programme

Half-sibs are brothers and sisters that share only a single parent in common. Regular programmes of half-sib matings are used when the same ultimate level of inbreeding is wanted, but a slower rate of increase is desired. A single generation of half-sib mating produces offspring with F = 12.5%, but the amount produced thereafter depends on the half-sib inbreeding programme that is used; F will reach 40–50% after four to five generations.

There are several types of regular half-sib inbreeding programmes. There are two matings each generation: one produces the male and the other produces the two half-sisters.

The inbreeding produced after the first generation in this programme is less than that produced by the smaller half-sib breeding family; the additional females in this breeding programme moderates the accumulation of inbreeding. The inbreeding programme illustrated herein, shows that the size of the family decreases each generation, but this is only because the illustration shows that each mating produces a single offspring that is used in the next generation. To keep family size from decreasing, one mating each generation that does not produce the male brood fish must produce two half-sisters.

The cost and complexity of half-sib inbreeding programmes depend on the programme that is used.

Half-sib mating programmes are used because large levels of inbreeding can be produced-but at a slower rate than is the case for parent-offspring or full-sib inbreeding programmes. This means regular half-sib inbreeding programmes take longer and are more expensive than full-sib and parent-offspring inbreeding programmes.

Double First Cousin Inbreeding Programme

Double first cousins are first cousins that are twice as related as regular first cousins because the parents that produced them are a pair of full-sibs that mated with a pair of full-sibs (often a pair of brothers that mate with a pair of sisters). A regular inbreeding programme of double first cousins is illustrated. The inbreeding produced by mating double first cousins is similar to that produced by regular systems of half-sib matings. The breeding programme illustrated may look complicated, but in reality it is a fairly simple regular mating scheme. Each inbred family in a breeding programme of double first cousin matings requires a minimum of four individuals per generation. Consequently, the cost of raising fish each generation is about twice that of a programme of full-sib matings.

Although this inbreeding programme produces levels of inbreeding that are similar to those produced by half-sib matings, it is less efficient and more expensive. When double first cousins are mated, no inbreeding will be produced until F_3-generation fish are produced, because the F_2 generation is the first generation in which double first cousins can be mated.

Chromosomal Manipulation

Chromosomal set manipulation can be used to produce highly inbred fish in a relatively short period. Individual fish with F = 100% can be produced in a single generation, while inbred lines where all fish have F = 100% can be produced in two generations.

Chromosomal set manipulation to produce inbred fish can be done in one of two basic ways, but regardless of the technique used, the fish that are produced have only a single parent. The first technique is to prevent the first mitotic division that occurs when the zygote nucleus and zygote itself divides to become a two-celled embryo. To create inbred fish, this technique is done with haploid zygotes. This technique is called either "mitotic gynogenesis" or "mitotic androgenesis," depending on the whether the haploid set of chromosomes of the zygote comes from the mother or from the father.

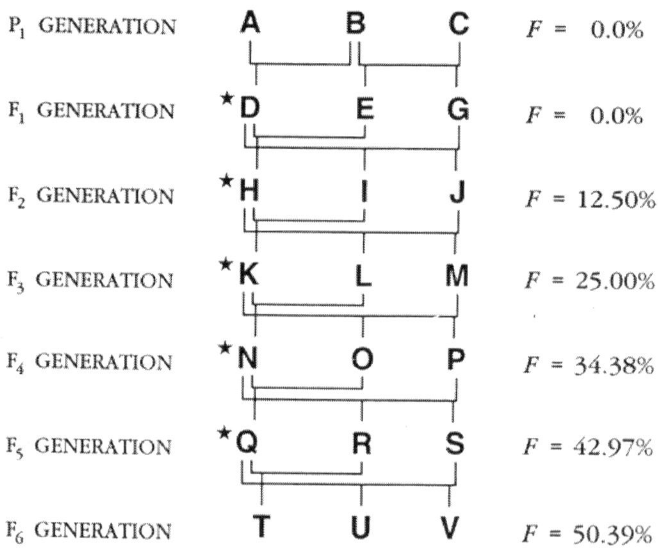

Figure: A regular inbreeding programme of half-sib matings. In this type of regular half-sib breeding programme, a single male each generation (marked with a star) is mated to two half-sisters. This is the simplest and least expensive type of half-sib inbreeding programme. Each inbred family is composed of three individuals per generation. There are two matings per generation: one produces the male and the other produces the two half-sisters. Five generations of half-sib matings and the inbreeding produced by this breeding programme are illustrated. The first generation of half-sib mating occurs in the F_1 generation, and their offspring (F_2 generation) is the first generation of inbred fish. Fish in the P_1 generation have F = 0% and are not related.

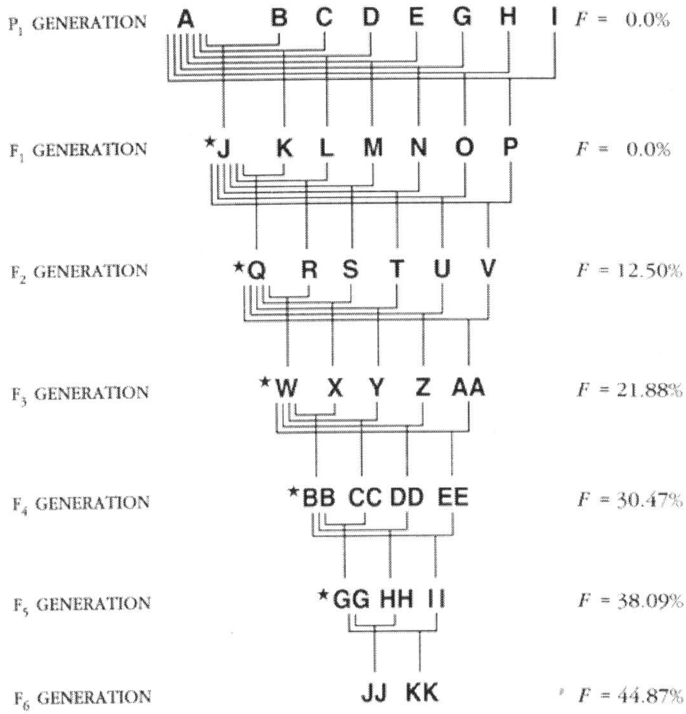

Figure: A regular inbreeding programme of half-sib matings. In this type of regular half-sib breeding programme, a single male each generation (marked with a star) is mated to many half-sisters. Five generations of half-sib matings and the inbreeding produced by this breeding programme are illustrated. The first generation of half-sib mating occurs in the F_1 generation, and their offspring (F_2 generation) is the first generation of inbred fish. Fish in the P_1 generation have F = 0% and are not related. This illustration shows that the size of the family decreases each generation. This is because each mating in the figure produces a single offspring that is used in the breeding programme. To keep family size from decreasing, one mating that does not produce the male must produce two half-sisters.

The second technique is to prevent equational division (second meiotic division) of the secondary oocyte (egg) after sperm penetration; this prevents the second polar body from leaving the egg. This technique is called "meiotic gynogenesis", because chromosomal set manipulation is accomplished by disrupting a meiotic division; it produces fish called "meiotic gynogens", because all chromosomes in the offspring come from the mother.

This technology requires highly skilled labour, and the methodologies have not been perfected. At present, these breeding programmes are important for some kinds of genetic research, but their practical use has not been quantified. Consequently, these breeding programmes should be done only by scientists who work at agribusinesses or research institutions that are capable of conducting sophisticated genetics experiments. The techniques and procedures needed to create the fish described in this section are species specific, so a detailed description for each important farmed species is beyond the scope of this manual.

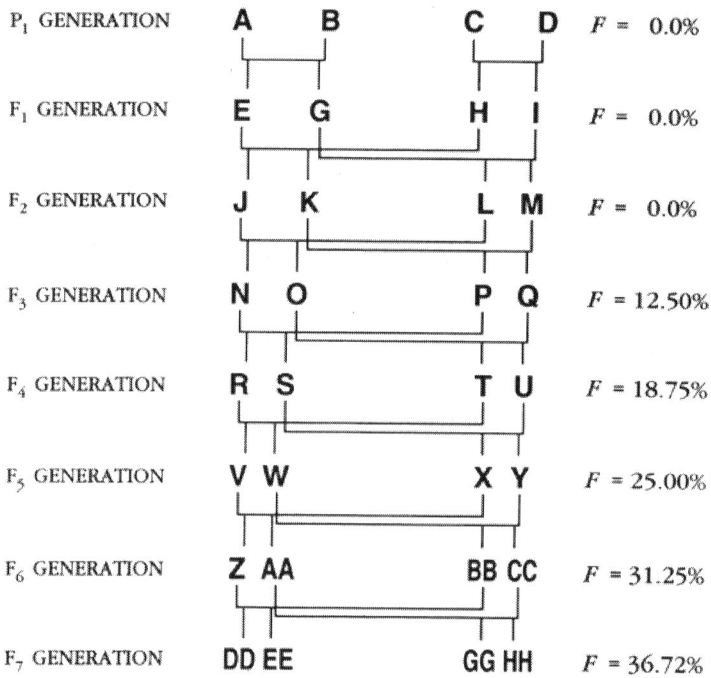

Figure: A regular inbreeding programme of double first cousin matings. Five generations of double first cousin matings and the inbreeding produced by the breeding programme are illustrated. The first generation of double first cousin matings occurs in the F_2 generation, and their offspring (F_3 generation) is the first generation of inbred fish. Fish in the P_1 generation have F = 0% and are not related.

Mitotic Gynogenesis

Mitotic gynogenesis can be used to create mitotic gynogens (all genes come from the mother), fish that are 100% inbred. The technique that is used to accomplish this with species that have the XY sex-determining system (females are XX and males are XY; virtually all aquacultured species have this system of sex determination) is outlined.

The first step in this breeding programme is the production of first-generation mitotic gynogens. Ultraviolet radiation is used to destroy the DNA (the genes) in sperm. The irradiated sperm are then used to activate eggs. An irradiated sperm cannot fertilize an egg because its genes have been destroyed. The activation causes the egg to undergo the equational division (second meiotic division) and to extrude the second polar body. The egg now contains only a haploid egg nucleus; this produces a haploid zygote (the zygote contains only a single chromosome (homologue) from each chromosome pair, and each chromosome comes from the mother, which is why they are called "gynogens").

When the haploid zygote undergoes first cleavage, a pressure or temperature shock is used to prevent the haploid zygote nucleus from dividing into two daughter nuclei. If the shock is timed perfectly, the haploid zygote nucleus has replicated its chromosomes so that each daughter nucleus will have a full and identical set of chromosomes, but the haploid zygote nucleus has not divided. By preventing first cleavage, the zygote remains a zygote, but the chromosome number of the zygote has doubled from the haploid state to the normal diploid state, which means that each chromosome occurs as a pair. Since mitosis (first cleavage is a mitotic cell division) produces two identical sets of chromosomes, each chromosome pair is composed of two identical chromosomes. Consequently, every gene comes from the mother, and every gene is homozygous; the mitotic diploid gynogen is 100% homozygous and 100% inbred.

This technique does not produce many viable fish. Since each fish is 100% homozygous, every detrimental recessive allele that reduces viability, produces an abnormality, or causes death will be expressed. However, mitotic gynogens that do survive are fish that are free from detrimental recessive alleles, which means they carry no recessive alleles that produce genetic defects.

If this is done with species that have the XY sex-determining system, each mitotic gynogen is a female. The only sex chromosome that exists in mitotic gynogens are those that were contributed by the mother, and females have only X sex chromosomes (all genetic material, including all sex chromosomes, in sperm were destroyed by UV irradiation). When the diploid chromosome number is restored by the shock, all fish go from X zygotes to XX zygotes, and all are females.

Each 100% homozygous mitotic gynogen is unique because there are thousands of genes. Consequently, each first-generation mitotic gynogen is a singular inbred line.

If first-generation mitotic gynogens are to be used in a breeding programme to create inbred lines, a second phase of gynogenesis followed by sex reversal is needed in order to produce the lines of 100% inbred fish, in which all fish within each inbred line are genetically identical. Each first-generation mitotic gynogen will be used to create a unique line of 100% homozygous and 100% inbred fish by utilising either of two possible types of gynogenesis. When the first-generation mitotic gynogens mature, their eggs are stripped, and either mitotic gynogenesis is repeated to produce second-generation gynogens or meiotic gynogenesis is used to create second generation gynogens. If meiotic gynogenesis is used, a shock is used to prevent the second polar body from leaving the activated but unfertilized eggs. Regardless of the technique used during phase 2 of this breeding programme, the second-generation gynogens are considered to be mitotic gynogens because that was the technique used during phase 1. Both techniques produce families (inbred lines) of identical fish that are 100% homozygous and 100% inbred but, in general, survival will be better if meiotic gynogenesis is used during phase 2.

The eggs and offspring from each first-generation mitotic gynogen must be isolated from all other second-generation eggs and offspring, because each first-generation mitotic gynogen is going to form a "cloned" line. The second round of gynogenesis produces hundreds of genetically identical copies of each first-generation mitotic gynogen (a form of cloning).

Half of the second-generation gynogens from each family are sex-reversed with anabolic androgens (steroid hormones) to produce XX sex-reversed males. The sex-reversed males are genetic females but phenotypic males. The fish that are not treated with hormones are raised normally. Within each family, the sex-reversed males and their sisters are genetically identical (genetically, they are all identical sisters); when they mate, they produce an inbred line of genetically identical fish that is 100% female, 100% homozygous, and 100% inbred. Sex-reversed males must be created every generation, because it is the only way males can be produced, and it is the only way each inbred line can be perpetuated without additional chromosomal manipulation.

If each inbred line is isolated from all other inbred lines, it will breed true and produce genetically identical 100% inbred fish. This breeding programme produces 100% inbred fish in one generation and 100% inbred fish that are capable of producing inbred lines of 100%

inbred fish in two generations. If this is done with species with the WZ sex-determining system (females are WZ and males are ZZ; the only major aquacultured species with this system of sex determination are: blue tilapia, Oreochromis aureus; Wami tilapia, O. urolepis; Japanese eel, Anguilla japonica), half the first-generation mitotic gynogens will be mitotic gynogenetic males (ZZ) and half will be mitotic gynogenetic superfemales (WW). Phase 2 of the breeding programme needed for mitotic gynogenetic WW superfemales is the same as that described previously in this section. ZZ mitotic gynogenetic males, on the other hand, must undergo mitotic androgenesis during phase 2.

Mitotic Androgenesis

Mitotic androgenesis can be used to produce mitotic androgens (all genes come from the father), fish that are 100% inbred. The technique that is used to produce mitotic androgens for species with the WZ sex-determining system is outlined.

In phase 1, eggs whose DNA has been destroyed by ultraviolet radiation are fertilized by normal sperm. This produces a haploid zygote, and all chromosomes come from the sperm (from the father), so these zygotes are called "mitotic androgens".

When the zygote undergoes first cleavage, a pressure or temperature shock is used to prevent nuclear division, which produces a diploid androgenetic zygote. In species with the WZ sex-determining system, males are homogametic (ZZ), so all sperm contain a Z chromosome. Consequently, all haploid zygotes have a Z sex chromosome. The disruption of first cleavage changes Z zygotes into ZZ zygotes, and all are males. As was the case with mitotic gynogens, all mitotic androgens are 100% homozygous and 100% inbred, but in this case all fish are males. Each androgen is a singular inbred line.

As was the case with mitotic gynogenesis, mitotic androgenesis does not produce many viable fish. Because each androgen is 100% homozygous, every detrimental recessive allele will be able to produce its phenotype. However, mitotic androgens that survive are free from detrimental recessive alleles.

Phase 2 of this breeding programme uses a second round of mitotic androgenesis, and half the offspring from each family are sex-reversed with anabolic estrogens to produce sex-reversed ZZ females. These females are genetic males but phenotypic females. Within each family, the sex-reversed females are mated to their genetically identical brothers to produce each inbred line; fish in each inbred line are genetically identical, 100% homozygous, and 100% inbred and all are males. As was the case with mitotic gynogens, each line is genetically unique.

Mitotic androgenesis can be done with species that have the XY sex-determining system. In this case, half the first-generation mitotic androgens will be XX androgenetic females and half will be YY androgenetic supermales. Second-generation androgens and inbred lines can be produced as follows: the phase 2 protocol needed for the YY androgenetic supermales.

Chapter 4

Meiotic Gynogenesis

Gynogenesis can be used to create another type of inbred fish-meiotic gynogens. This type of chromosomal manipulation is easier than mitotic gynogenesis, and meiotic gynogens have a higher survival rate than mitotic gynogens because they have less inbreeding. Meiotic gynogenesis is less useful in producing inbred lines because it is difficult to accurately predict the exact amount of inbreeding produced, and the inbreeding produced each generation is quite variable. Since regular systems of inbreeding are most useful when they produce reliable and predictable amounts of inbreeding, meiotic gynogenesis is less useful than mitotic gynogenesis for producing inbred lines. However, one to three generations of meiotic gynogenesis can be used to produce highly inbred fish.

Ultraviolet irradiated sperm are used to activate eggs. A pressure or temperature shock is applied shortly after activation to prevent the second polar body from leaving the egg during the equational meiotic division. The shock produces an activated egg that has a haploid egg nucleus and a haploid second polar body nucleus. These haploid nuclei fuse, and the egg "fertilizes" itself to produce a meiotic gynogenetic diploid zygote.

At first glance, it appears as if meiotic gynogenesis is a form of self-fertilization, which should theoretically produce fish with F = 50% in the first generation, F = 75% in the second generation, and F >99% in the seventh generation. However, this isn't a form of self-fertilization because the genes in a haploid egg nucleus and the haploid second polar body nucleus are not truly random with respect to each other, as is the case with two randomly chosen haploid egg nuclei. A haploid egg nucleus and the haploid second polar body nucleus are created after the reduction division. Two randomly chosen haploid egg nuclei differ from each other in that independent assortment randomly distributed the maternal and paternal homologues of each chromosome pair. An egg haploid nucleus and the second polar body haploid nucleus have already undergone independent assortment, so both receive a replicated homologue from each pair of sister chromatids; consequently, they differ less from each other in that respect than two randomly chosen egg nuclei. When the egg haploid nucleus and the haploid second polar body nuclei fuse, the sister chromatids that were separated during the equational division are reunited. These sister chromatids tend to be homozygous with respect for genes near the centromere and heterozygous for genes further away due to crossing over. Because of this, the exact amount of inbreeding produced by one generation of meiotic gynogenesis is difficult to predict.

Meiotic Gynogenesis Inbreeding

To further complicate matters, It is impossible to accurately predict the amount of inbreeding that will be produced by subsequent generations of meiotic gynogenesis, because the inbreeding will be a function of crossing over frequencies. Crossing over tends to occur at particular locations along each chromosome, so heterozygous loci tend to remain heterozygous for many generations. Because heterozygous loci will tend to remain heterozygous, full-sib mating will eventually produce greater levels of inbreeding than meiotic gynogenesis. Additionally, the inbreeding that is produced each generation will be different for each species and each population due to differences in crossing over frequencies.

The only way to quantify the inbreeding produced by meiotic gynogenesis is to examine the mothers and their offspring electrophoretically. Heterozygous loci in diploid meiotic gynogens that are produced from mothers that are heterozygous at those loci are heterozygous because of crossing over. The percentage of such loci that are heterozygous in the offspring can be used to determine crossing over frequency; once this is ascertained, it can be used to determine inbreeding for each meiotic gynogen. Studies with fish have shown that one generation of meiotic gynogenesis produced F = 55% to 79% (F = 79% is approximately that produced by seven generations of full-sib matings).

Although meiotic gynogenesis produces less inbreeding than mitotic gynogenesis, it can be used to quickly produce highly inbred fish. However, each meiotic gynogen will have a different level of inbreeding.

Most aquaculturists have been told that inbreeding causes problems and that it should be avoided at all costs. This is true only if inbreeding is undesired, because inbreeding is a powerful and valuable breeding technique that animal and plant breeders have used for years to produce superior brood stock and to produce superior plants and animals for grow-out. Any farmer or hatchery manager who wants to manage the genetic aspects of his population should know when and how to use inbreeding to improve productivity and profits, as well as when it should be avoided so that inbreeding depression does not ruin productivity and profits.

Inbreeding is used to create new breeds that breed true to form. Linebreeding can be used to bring a population up to the level exhibited by a superior animal and to increase a superior animal's contribution to his descendants. The use of herd bulls is a type of inbreeding that can be used to quickly improve a population. The creation of inbred families can improve the results of between-family selection when a breeder wants to improve a phenotype that has a low h^2. Finally, the creation of inbred lines is a way to increase hybrid vigour in crossbreeding programmes and to produce outstanding F_1 hybrids for grow-out.

A farmer can use inbred lines to improve selection and crossbreeding programmes only if he knows how to produce inbred lines efficiently. This means that he must know how to conduct regular systems of inbreeding so that he knows what matings must be made, generation after generation, in order to produce predictable levels of inbreeding. Several types of regular inbreeding programmes can be used to produce inbred lines. Parent-offspring and full-sib inbreeding programmes are the simplest and cheapest, and both quickly produce high levels of inbreeding. Full-sib inbreeding programmes are usually used because it produces large amounts of inbreeding, and it is rather easy to make this type of mating generation after generation. Half-sib and double first cousin inbreeding programmes are slightly more complicated and somewhat more expensive. Although they ultimately produce the same amount of inbreeding as parent-offspring and full-sib inbreeding programmes, they produce it at a slower rate.

Chromosomal manipulation can be used to quickly produce highly inbred lines of fish. One generation of mitotic gynogenesis or mitotic androgenesis will produce fish that are 100% homozygous and 100% inbred. A second generation of chromosome set manipulation is needed to produce 100% inbred fish that are capable of reproducing. One generation of meiotic gynogenesis will produce fish that have large, but unknown, levels of inbreeding; the amount of inbreeding produced by meiotic gynogenesis is variable and depends on crossing over frequencies. The use of chromosomal manipulation to produce inbred lines should be done only by scientists at large agribusinesses or at research stations.

Preventing Inbreeding Depression and Loss of Genetic Variance in Hatchery Populations

Although inbreeding is a valuable breeding technique and one that can be combined with selection to improve response when h^2 is small or with crossbreeding to produce animals and plants that farmers grow for market, unplanned and uncontrolled inbreeding often leads to inbreeding depression-decreases in growth rate, fecundity, and survival. The problems that are produced by uncontrolled inbreeding are usually accompanied by loss of alleles via genetic drift.

When a farmer or hatchery manager chooses a management programme, either he elects to improve productivity by utilising selection, crossbreeding, inbreeding, or a combination of these programmes; or he decides to forego genetic improvements and will not use any breeding programme. If a farmer uses a breeding programme, it is inevitable that he will produce inbred fish, and the level of inbreeding will increase over time. Additionally, genetic drift will alter gene frequencies. However, inbreeding and genetic drift are not as important under these circumstances as they are when no breeding programme will be utilised. The reasons why inbreeding and genetic drift are not as important under these circumstances.

When no breeding programme is used to improve the population or to produce genetically improved fish, the only genetic goals should be: to prevent or minimise inbreeding, as well as to prevent genetic drift from reducing genetic variance and to minimise the loss of alleles. When a breeding programme is used, genetic changes are planned and desired because they will lead to improved growth rate, etc. When no breeding programme is conducted, genetic changes are unplanned and random, and those caused by inbreeding and genetic drift can ruin the population.

Consequently, when a farmer or hatchery manager does not incorporate a breeding programme into his management plan, it is advisable to try and prevent inbreeding from reaching levels that depress productivity and profits and to prevent genetic drift from robbing the population of potentially valuable alleles.

This chapter will describe ways to prevent inbreeding from accumulating to levels that cause inbreeding depression when farmers and hatchery managers do not use selection or other breeding programmes, both when fish can be marked and identified and when fish are not marked. Since most farmers and hatchery managers do not mark fish, this usually means that N_e must be managed to minimise inbreeding. The management of N_e to prevent genetic drift from robbing the population of valuable alleles will also be described.

Managing N_e to prevent inbreeding- and genetic drift-related problems means managing N_e at a pre-determined size every generation. Maintaining N_e at a constant size is difficult, because N_e can decline due to diseases, poor spawning season, etc. If N_e declines, the population goes through what is called a "bottleneck." Bottlenecks accelerate the accumulation of inbreeding and magnify genetic drift. Preventing bottlenecks, especially during acquisition of the population, may be the most important aspect of brood stock management.

Preventing Inbreeding Depression

The first step in managing a hatchery population to prevent inbreeding depression is to decide what level of inbreeding causes problems. Unfortunately, there is no answer to this seemingly simple question.

There have been relatively few inbreeding studies with fish. Some of those that have been conducted were done on hatchery fish which were probably already inbred. Even if a major study were to be conducted for an important species, such as common carp, Cyprinus carpio, or Nile tilapia, the inbreeding values that cause inbreeding depression and the level at which inbreeding depression become economically significant might not be the same for all hatchery populations of that species. Studies with other animals have shown that any level of inbreeding can cause inbreeding depression, but the level at which it becomes significant varies from study to study and is often different for different phenotypes. Consequently, no universally undesirable value of F will ever exist for fish. It will be different for different species and for different populations within a species, and it will also be different for different phenotypes.

Because no universally undesirable value of F will ever exist, farmers and hatchery managers have to make intelligent "guesstimates" about what levels of inbreeding they want to avoid. The inbreeding study with rainbow trout that was illustrated, showed that inbreeding depression became a serious problem when Fe"18%. However, F = 12.5% also produced some inbreeding depression. The effects of F <12.5% were not evaluated, so the effects of mild levels are not known. Most inbreeding studies with fish have examined the effects of Fe"25% (F = 25% is one generation of brother-sister matings).

Because there have been so few inbreeding studies in fish, because the effects of mild levels have not been investigated, and because no universally undesirable value of F is likely to be found, farmers and hatchery managers must decide what level of risk they are willing to accept. Those who want little risk should try to prevent inbreeding from exceeding 5%. Those who are willing to accept more risk should try to prevent inbreeding from exceeding 10%. In this manual, "risk" is defined as potential problems due to inbreeding and genetic drift; low levels of risk (smaller levels of inbreeding are desired) require more management, while high levels of risk (larger levels of inbreeding are acceptable) require less management.

These inbreeding values are not carved in stone. They are simply suggested values for populations where there is no selection to improve growth rate or other production phenotypes. Farmers and hatchery managers can substitute any value they choose. Small levels of inbreeding are more difficult and more costly to prevent, but efforts to prevent small levels will produce populations with fewer genetic problems. Large levels require less management and are less expensive, but choosing such levels may mean that the population has to be replaced when inbreeding depression rears its ugly head.

If Fish are Marked

If each fish receives an individual mark, tag, or brand, pedigrees can be developed, which means that individual inbreeding values can be determined. If all fish are marked, inbreeding can be prevented in small populations. The way this is accomplished is: pedigrees are created, and relatives are not allowed to mate. One way to help ensure this is to construct a table of covariance values for all brood fish. If the Cov_{BI} value for two brood fish is 0.0, those fish are allowed to mate, because the offspring that will be produced will have F = 0%.

If a farmer wants to allow some inbreeding but wants to keep it below a certain level, he can use the table of covariance values to determine the matings that will produce inbreeding values within his desired range (0% to the level that has been chosen). To do this, the farmer allows all matings where Cov_{BI}d"2 (maximum desired level of inbreeding). For example, if a farmer wants to keep F ≤5%, he allows all matings where Cov_{BI} ranges from 0.0 to 0.10, and prevents those where Cov_{BI}>0.10.

If a farmer marks his fish and maintains pedigrees, he can use the information, to determine which matings will enable him to prevent undesired levels of inbreeding. Some types of consanguineous matings produce moderate amounts of inbreeding, while others produce virtually no inbreeding. If very low levels of inbreeding are desired, a farmer can restrict consanguineous matings to those less related than first cousins. If second cousins are mated generation after generation, inbreeding will never exceed 2%. For practical purposes, a regular inbreeding programme of second cousin matings is an effective way to prevent inbreeding depression. Unfortunately, a regular inbreeding programme of second cousin matings is fairly complicated.

If Fish are not Marked

If fish are not marked, individual inbreeding values cannot be determined, so a farmer must try to prevent the average inbreeding value from exceeding his desired level. This means that the farmer must manage his population's N_e.

Two important decisions must be made before a farmer or hatchery manager can manage N_e to prevent inbreeding from exceeding the desired maximum level: The first is to decide what the maximum inbreeding value will be. As was described earlier in this chapter, F = 5% is proposed for conservative farmers and hatchery managers who want to take little risk, while F = 10% is proposed for those who are willing to accept more risk. However, each farmer or hatchery manager is free to choose his level of risk and to customize the maximum level of inbreeding for his population.

The second decision a farmer or hatchery manager must make is to determine the number of generations that will be incorporated into the work plan before exceeding the level of inbreeding that has been chosen. In this exercise, a generation is the replacement of brood fish by their offspring, and generations are not allowed to overlap (the mixing of generations).

Once these two decisions have been made, two simple procedures are used to determine the N_e that is needed to achieve a farmer's goal:

Step 1: Once the desired maximum level of inbreeding and the number of generations that will be incorporated into the work plan have been determined, the amount of inbreeding per generation that is needed to produce the maximum level in that number of generations must be determined. The only sensible way N_e can be managed to achieve such a goal is to maintain a constant N_e, which will produce a constant amount of inbreeding. The reasons for this are: the math needed to determine the N_e's is far simpler, and it is far easier to try and maintain a constant N_e than a fluctuating one. The formula needed to determine this value is:

$$F / generation = \frac{total\ F\ desired}{number\ of\ generations}$$

Step 2: Once the inbreeding per generation has been determined, the N_e that is needed to produce that level of inbreeding can be calculated by using the following formula:

$$F = \frac{1}{2Ne}$$

For example, if a farmer decides that he wants to keep inbreeding from exceeding 5% and he does not want inbreeding to reach 5% until he produces the 10th generation, the N_e that is needed can be determined as follows:

Step 1. Calculate the inbreeding per generation that will produce F = 5% (0.05) after 10 generations:

$$F / generation = \frac{total\ F\ desired}{number\ of\ generation}$$

$$F / generation = \frac{0.05}{10}$$

F/generation = 0.005

Thus, if F = 0.005/generation, it will reach 5% (0.05) when the 10th generation is produced.

Step 2. Calculate the N_e per generation that will produce F = 0.005/generation:

$$F = \frac{1}{2Ne}$$

$$0.005 = \frac{1}{2Ne}$$

$$0.005(2) = \frac{1}{Ne}$$

$$Ne = \frac{1}{0.005(2)}$$

N_e=100

The farmer can achieve his goal of preventing inbreeding from exceeding 5% until he produces the 10th generation if he maintains an N_e = 100 for all 10 generations. If N_e = 100 for nine of the 10 generations but drops below 100 for a single generation, he will not be able to achieve his goal. It is important to note that this procedure determines a minimum N_e, not the number brood fish that must spawned. The two are usually not the same. The only time they are the same is if a 1:1 sex ratio is used, if spawning success is 100%, and if each mating produces viable offspring. In this occurs, a farmer simply divides N_e by 2 to determine the number of males and females that must be spawned. In this example, a farmer would have to spawn 50 males and 50 females every generation to achieve his goal. If the farmer uses a skewed sex ratio (usually more females than males), he will need more than 100 brood fish. The exact number he would need will be determined by the rarer sex.

If reproductive success is not 100%, a farmer must save and maintain more brood fish than the N_e that he is trying to achieve. The number of brood fish that must be maintained is determined by the typical spawning success rate and by the percentage of matings that produce viable offspring. For example, if a farmer wants an N_e = 100, if 75% of the fish spawn, if 95% of those that spawn produce viable offspring, and if survival from the time brood fish are set aside until they are spawned is 70%, the farmer determines the number of brood fish that he needs in the following manner:

Step 1. Determine how many males need to spawned. N_e = 100; it will be assumed that the sex ratio will be equal (1:1), so 50 males and 50 females are needed.

Step 2. Determine how many fish are needed if some matings produce no viable offspring. In this example, 95% of the males produce viable offspring, so the number of males that are needed is:

$$Number\ of\ males = \frac{number\ desired}{percent\ producing\ viable\ offspring}$$

$$Number\ of\ males = \frac{50}{0.59}$$

$$Number\ of\ males = 52.63$$

Since the sex ratio in this example is 1:1, the number of males and number of females that are needed are the same; consequently, the number of females needed = 52.63.

Step 3. Determine how many male brood fish are needed if spawning success is <100%. Spawning success is 75%, and 52.63 males will spawn, so the number of male brood fish that will be needed is:

$$Number\ of\ males = \frac{number\ that\ spawn}{spawning\ success}$$

$$Number\ of\ males = \frac{52.63}{0.75}$$

Number of males = 70.17

Since the number of males and females are equal, the number of females needed = 70.17.

Step 4. Determine how many fish must be saved and set aside if survival is <100%. Survival from the time fish are set aside until they are spawned is 70%; 70.17 males are needed, so the number that must be set aside is:

$$Number\ set\ aside = \frac{number\ of\ male\ sneeded\ at\ spawing}{surival}$$

$$Number\ set\ aside = \frac{70.17}{0.70}$$

Number set aside = 100.24

Since the number of males and females are equal, the number of females needed = 100.24.

Because you cannot have two-tenths of a brood fish, the number is rounded up to 101. Consequently, in this example, the farmer would need to save 101 male and 101 female brood fish to produce an N_e of 100. The same numbers of males and females are needed in this example, because a 1:1 sex ratio was used and each male and each female were mated only once. If a skewed sex ratio is used or if multiple matings are used, the arithmetic becomes more complicated, but the process that was just described can be used to determine how many brood fish must be used.

Effective breeding numbers that are needed to keep inbreeding from exceeding 1–15% for 1–100 generations. When calculating the N_e that was needed to achieve a desired level of inbreeding, the number that was determined was often not a whole number. When this occurred, the number was rounded to the next higher whole number. This was done because you cannot have a fraction of a fish, and rounding to the next higher whole number ensured that F would not exceed the desired level. For example, an N_e of 83.3 was rounded up to 84. If the number were rounded down, it would fall below the N_e that was calculated, which would mean that the goal could not be achieved.

Preventing Genetic Drift

As was mentioned earlier, uncontrolled inbreeding and genetic drift are twin evils that occur in closed hatchery populations. Because they are conjoined twins and because both are determined by the population's N_e, it is important to know how to manage and control both. In many cases, managing the population to prevent one of these problems will prevent the other. In cases where this does not occur, a slight alteration in management goals (the desired N_e) will prevent both problems from adversely affecting productivity and profits.

It is not possible to prevent genetic drift. Genetic drift is simply random changes in gene frequency that occur because of sampling error, and the only way to prevent sampling error is to have an infinitely large population which is impossible. In aquaculture, sampling error is choice of brood fish that are allowed to spawn and produce the next generation or the acquisition of fish from another fish farm or the wild. Genetic drift causes problems when the random changes in gene frequency cause some alleles to be lost (the frequency goes to 0.0 or 0%). Once lost, these alleles can be regained only via mutation (which is rare) or by the acquisition of new brood fish. Rare alleles (frequencies <0.01) will be lost more easily than common ones, but common ones can also be lost. The probability of losing an allele is related to its frequency and the population's N_e.

Consequently, a hatchery manager who wants to manage his hatchery population in order to prevent unwanted inbreeding from ruining productivity or profits must also manage the population to prevent genetic drift from robbing the population of genetic variance (loss of alleles).

Managing a population to control genetic drift means that a farmer or hatchery manager has to decide how much genetic drift is acceptable. Since genetic drift cannot be stopped, the question has to be modified. For management purposes, the question becomes: How large does N_e have to be to prevent the loss of alleles?

To answer this question, two decisions must be made. The first is: How valuable are the rare alleles; i.e., what is the frequency of the alleles that will be saved? The second is: What guarantee of saving these alleles is desired? The reason these decisions must be made is because the only way the loss of alleles can be prevented and the only way a 100% guarantee can be given that alleles were not lost via genetic drift is to have an infinitely large population. Hatchery populations are usually small, so the only way to manage them and prevent genetic drift from causing problems is to make a compromise between what is ideal (no changes in gene frequency and no loss of alleles) and what is achievable (determined by the above decisions).

Population biologists generally assume that if a gene has more than one allele and if the frequency of two alleles are e"0.01, the gene is polymorphic, which means that for management purposes two or more alleles exist for that gene in the population. Alleles whose frequencies are <0.01 do not contribute to polymorphism. This does not mean they are unimportant; they do contribute to genetic variance, but they are often considered too rare to manage for practical biological conservation.

Consequently, a farmer or hatchery manager who wants to preserve as much genetic variance as possible and wants to accept little risk would choose to save alleles whose frequencies are 0.01. Farmers who feel that rare alleles are not that important for farming purposes would choose to save more common alleles. They might choose to save alleles whose frequencies are, say, 0.05. Alleles that are rarer than 0.05 are probably not that important in fish farming. If they were, domestication selection would have increased their frequency dramatically.

The frequency that is chosen is determined by the farmer's or hatchery manager's goals. There is no single correct value. However, hatchery managers who are raising fish that will be stocked in lakes and rivers should try and preserve as much genetic variation as possible and should try and prevent any changes in the genetic make-up of the population. This means saving alleles whose frequencies are 0.01–0.001. Farmers who raise food fish do not need to be as zealous. If they wish to conserve the population's genetic variance, they only need to save alleles whose frequencies are 0.05-0.01; however, many fish farmers should not even worry about the effects of genetic drift.

As was mentioned earlier, it is not possible to manage an N_e to produce a 100% guarantee of saving an allele, so practical workable guarantees must be chosen. In biology, two values are routinely used: 95% and 99%. These values come from statistics: when biologists evaluate experiments statistically, they typically use 0.01 and 0.05 as probability levels. The guarantee of saving an allele is: 1.0 - the probability of losing the allele.

A final decision must be made before the N_e can be managed to prevent the loss of rare alleles: How many generations will be incorporated into the management plan; i.e., how many generations will be produced until the guarantee of keeping the alleles is lowered to the desired level?

Once these decisions have been made, the N_e that is needed to prevent genetic drift from robbing the population of rare alleles is determined by using the following formula:

$$P = (1.0 - q)^{2N_e}$$

Where: P is the probability of losing the allele in a single random sample (the fish that mate to produce the next generation or the transfer of fish from one hatchery to another); and q is the frequency of the rare allele that is to be saved (q = 0.01 or 0.05 or whatever value is chosen).

Given: You want to save alleles whose frequencies are 0.01 (q), and you want a 95% guarantee of saving the allele (P = 0.05) after 10 generations (when the 10th generation is produced). What constant N_e is needed to achieve this goal?

Step 1: Calculate the guarantee per generation that is needed to produce a 95% guarantee after 10 generations:

$$0.95 = (\text{guarantee/generation})^{10}$$

$$\text{guarantee/generation} = (0.95)^{1/10}$$

$$\text{guarantee/generation} = 0.994883803$$

$(0.95)^{1/10}$ can be determined by using the "y^x" button on a hand-held calculator.

Step 2: Calculate the probability of losing the allele per generation:

Probability of losing the allele/generation = 1.0 - guarantee of saving it /generation

Probability of losing the allele/generation = 1.0 - 0.994883803

Probability of losing the allele/generation = 0.005116196882

Step 3: Calculate the N_e that is needed to produce a P = 0.005116196882 when q = 0.01 (the frequencies of the alleles you are trying to save):

$$P = (1.0 - q)^{2N_e}$$

$$0.005116196882 = (1.0 - 0.01)^{2N_e}$$

To determine N_e, the formula must be converted to logarithms. This can be done by using the "log" button on a hand-held calculator.

$$\log 0.005116196882 = \log(0.99)^{2N_e}$$

$$\log 0.005116196882 = (2N_e)(\log 0.99)$$

$$\frac{\log 0.0051161196882}{\log 0.99} = 2Ne$$

$$-\frac{2.291052751}{-0.004364805402} = 2Ne$$

$$524.89 = 2N_e$$

$$\frac{524.89}{2} = Ne$$

$$262.45 = N_e$$

N_e is rounded to the next higher whole number, so an N_e of 263 is needed every generation in order to produce a 95% guarantee of saving an allele whose frequency is 0.01 when the 10th generation is produced.

Bottlenecks

All of the recommendations that were made in the previous section were based on maintaining minimum constant N_e's. Unfortunately, it is difficult to maintain a constant N_e generation after generation. Anyone who has ever managed a population of fish knows all too well that it is difficult to maintain a steady-state population. Many factors conspire to occasionally reduce population size. Sudden drastic decreases in population size are called "bottlenecks." The genetic effects of bottlenecks can be devastating and can have long-term repercussions.

The mean N_e over a series of generations is the harmonic mean, not the simple arithmetic mean. Consequently, the generation with the smallest N_e has a disproportionate influence on the average value. This means that a bottleneck can dramatically lower mean N_e, which in turn will dramatically increase inbreeding and genetic drift.

For example, if a farmer wants to maintain a constant N_e of 100 for 10 generations but experiences a bottleneck of 20 in generation 6, the mean N_e that he produced is:

$$\frac{1}{Ne\,mean} = \frac{1}{t}\left(\frac{1}{Ne1} + \frac{1}{Ne2} + \ldots + \frac{1}{Net}\right)$$

$$\frac{1}{Ne\,mean} = \frac{1}{10}\left(\frac{1}{100} + \frac{1}{100} + \frac{1}{100} + \frac{1}{100} + \frac{1}{100} + \frac{1}{20} + \frac{1}{100} + \frac{1}{100} + \frac{1}{100}\right.$$

$$\frac{1}{Nemean} = 0.1(0.01 + 0.01 + 0.01 + 0.01 + 0.05 + 0.01 + 0.01\text{-}$$

$$\frac{1}{Ne\,mean} = 0.014$$

$N_{e\ mean}$ can be determined by using the "1/x" button on a hand-held calculator.

$$Ne\,mean = 71.4$$

The arithmetic average for this series of N_e's is 92, so mean N_e is 22% smaller than would be expected. If there were two bottlenecks of 20, mean N_e would have dropped to only 55.6.

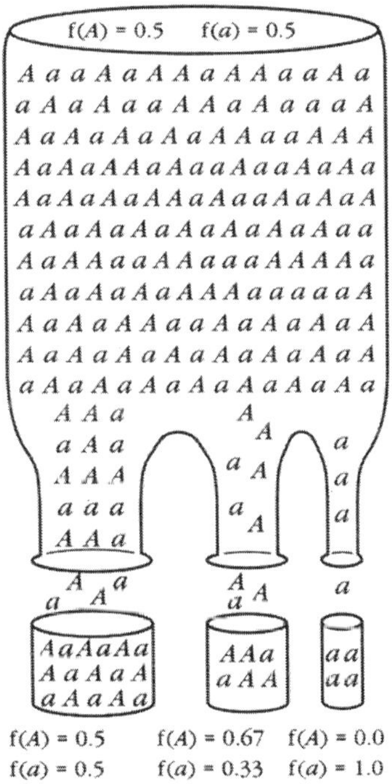

Figure: Effects of bottlenecks on gene frequency. Since genetic drift and the probability of losing alleles are inversely related to effective breeding number (N_e), the size of a bottleneck determines the amount of genetic change. The frequencies of the A and a alleles in the population are both 0.5. As the bottleneck (N_e) becomes smaller, genetic drift alters gene frequency until at an extreme, the frequency of an allele goes to zero.

Bottlenecks can have severe and long-lasting effects on inbreeding, because once inbreeding occurs, it lowers future N_e's, as was described. The impact that bottlenecks have on average inbreeding depends on the size of the bottleneck and how many there are. A population which is poorly managed and which experiences multiple bottlenecks where the population is reduced to relatively few males and/or females will be quite inbred. On the other hand, a well-managed population that experiences a single bottleneck will be far less affected. A single bottleneck will produce an immediate increase in inbreeding and the average inbreeding may increase for a generation or two (the amount depends on the size of the bottleneck), but if the population is properly managed to minimise inbreeding before and after the bottleneck, the average inbreeding will plateau a few generations after the bottleneck, and the damage due to inbreeding might not be severe.

The effects that bottlenecks can have on genetic drift are far more devastating. The probability of losing an allele is inversely related to N_e, so the odds of losing the allele increase as N_e decreases. Bottlenecks increase the probability dramatically. For example, if a farmer is trying to maintain a constant N_e of 344 for 10 generations in order to produce a guarantee of 99% for keeping an allele whose frequency is 0.01 and there is a bottleneck of 25 in generation 8, the probability of losing the allele when the 10th generation is produced is determined as follows:

Step 1: The probability of losing the allele each generation must be determined.

Step 2: The guarantee of keeping the allele each generation must be determined (1.0 - probability of losing the allele).

The probabilities of losing the allele and the guarantees of keeping the allele for all 10 generations are:

N_e	Step 1: Probability of losing allele		Step 2: Guarantee of keeping allele
	$(P = (1.0 - q)^{2N_e})$		(1.0 - P)
	344 $P = (0.99)^{2(344)}$	= 0.0009931477	0.999006852
	344 $P = (0.99)^{2(344)}$	= 0.0009931477	0.999006852
	344 $P = (0.99)^{2(344)}$	= 0.0009931477	0.999006852
	344 $P = (0.99)^{2(344)}$	= 0.0009931477	0.999006852
	344 $P = (0.99)^{2(344)}$	= 0.0009931477	0.999006852
	344 $P = (0.99)^{2(344)}$	= 0.0009931477	0.999006852
	344 $P = (0.99)^{2(344)}$	= 0.0009931477	0.999006852
	25 $P = (0.99)^{2(25)}$	= 0.605006067	0.394993932
	344 $P = (0.99)^{2(344)}$	= 0.0009931477	0.999006852
	344 $P = (0.99)^{2(344)}$	= 0.0009931477	0.999006852

Step 3: The guarantee of keeping the allele for 10 generations must be determined. It is the product of the guarantee of keeping the allele for each generation; i.e., the guarantees for each generation are

multiplied to determine the guarantee after the 10th generation is produced:

Guarantee after 10 generations	=	(0.999006852)(0.999006852)(0.999006852)(0.999006852)(0.999006852)(0.999006852)(0.999006852)(0.394993932)(0.999006852)(0.999006852)
Guarantee after 10 generations	=	0.3915

Step 4: The probability of losing the allele after 10 generations must be determined. It is 1.0 - the guarantee of keeping the allele:

Probability of losing allele = 1.0 - Guarantee of keeping allele

Probability of losing allele = 1.0 - 0.3915

Probability of losing allele = 0.6085

The bottleneck of 25 at generation 8 increased the probability of losing the allele in the 10th generation from 0.01 (which was the goal) to 0.6085.

The overall effect of a single bottleneck on genetic drift is far more severe than it is on average inbreeding, and its occurrence can prevent a farmer or hatchery manager from achieving his goals, even if he has done an outstanding job for many generations. This is because once an allele is lost, it can be recovered only by mutation or the introduction of new brood stock. Consequently, prevention of bottlenecks should be the major brood stock management goal if the genetic goal is to minimise the detrimental effects of inbreeding and genetic drift.

Effective Breeding Number

Since inbreeding and genetic drift are controlled by the population's N_e, it should be obvious that managing N_e is the most effective way to control or to prevent these twin problems from ruining productivity and profits. Minimum constant N_e's need to be maintained to prevent inbreeding from exceeding desired levels or to produce desired guarantees of keeping alleles with certain frequencies.

However, there are some management techniques that can be used to "artificially" increase N_e or to manage the brood fish to "stretch" generations. These techniques lengthen the time frame before inbreeding and genetic drift cause problems. By incorporating some of these management techniques, smaller numbers of brood fish can be used; smaller N_e's can be used; or the brood fish that are spawned can be managed more effectively to prevent genetic degradation of a population.

Stretching Generations

The simplest and least expensive technique that can be used to increase the time frame until inbreeding- and genetic drift-related problems cause trouble is to stretch the generation interval. This is not always possible, since some fish die when they spawn. But when it is possible, this technique is very effective. In this manual, a generation is defined as the replacement of brood fish by their offspring. Generations are not allowed to overlap. The definition of "generation" does not contain a specific time frame in terms of years. So if the normal time it takes for a fish to go from egg to brood fish and then for that fish to spawn and be discarded so that it can be replaced by its offspring is 2 years, the generation interval for that species is 2 years. If the time interval for another species is 4 years, the generation interval for that species is 4 years.

But these time intervals can be stretched if brood fish are allowed to spawn for an extra year or two before they are replaced by their offspring. The longer a generation can be stretched, the longer it will take for inbreeding-and genetic drift-related problems to become evident. For example, let's assume that the generation interval for the species a farmer cultures is 2 years. Let's also assume that he wants to keep inbreeding below 5% for 10 generations (20 years). To do this, he needs an N_e of 100 per generation. However, if he can stretch the generation interval so that it doubles to 4 years, he can use an N_e of only 50 to produce the same amount of inbreeding (5%) at the end of 20 years, because he will produce only five generations during this time span. Consequently, a farmer can maintain and spawn fewer fish by stretching the generations.

The major constraint with this type of brood stock management is survival. If survival of brood fish declines during this "extra" period, N_e of the population will decline. However, if survival of brood fish is not adversely affected by the extra stretched interval, this technique is an easy way of controlling inbreeding. One way of compensating for mortalities is to maintain additional brood fish and to use them to replace those that die.

Pedigreed Mating

Most farmers or hatchery managers use what is called "random mating" when they spawn their fish. Random mating is a breeding protocol where fish are mated without regard to phenotypic value. The only time random mating is not used is when selective breeding or crossbreeding programmes are implemented to improve growth rate or other phenotypes.

The N_e of a population can be "artificially" increased by using a breeding programme called "pedigreed mating." Pedigreed mating differs from random mating in that each female leaves one daughter and each male leaves one son to be used as brood fish in the following generation. In actuality, each male and each female can leave more than one descendant, but all fish must leave the same number. This means that a male that spawns with five females will leave the same number of sons as a male that spawns with only one female. They will leave unequal numbers of daughters, but they will leave the same number of sons. The sons and daughters that become brood fish in the following generation must be chosen randomly from each family. No fish can be chosen because it has a specific phenotype or is a certain weight; if that is done, some form of selection is occurring.

When pedigreed mating is used, the N_e of the population increases because the genetic variance is artificially increased by ensuring that each brood fish is represented in the next generation. When pedigreed mating is used, N_e is determined by using the following formula:

$$Ne = \frac{16(Number\ of\ males)(number\ of\ females)}{3(number\ of\ males) + number\ of\ females}$$

if there are more males,

or

$$Ne = \frac{16(Number\ of\ males)(number\ of\ females)}{3(number\ of\ females) + number\ of\ femles}$$

If there are more females.

If equal numbers of males and females are spawned, either formula can be used.

For example, if a farmer spawns 50 females and 30 males using random mating, his N_e is:

$$Ne = \frac{4(Number\ of\ males)(number\ of\ females)}{(number\ of\ females) + number\ of\ males}$$

$$Ne = \frac{16(30)(50)}{3(50) + 30}$$

N_e = 75

If he uses pedigreed mating, the N_e increases to:

N_e= 133.3

Pedigreed mating increased N_e by 77.8% in the above example, and the inbreeding that would be produced would decline from 0.67% to 0.38%. The only difficulty in using pedigreed mating is that you have to be able to mark the fish or be able to grow each family in an isolated pond or tank so that individual families can be identified.

Altering Spawning Techniques

There are several practical ways that spawning can be modified in order to increase N_e and thus reduce inbreeding and the effects of genetic drift.

Increasing the Spawning Population

— One way to reduce inbreeding and the effects of genetic drift is to spawn more fish than are needed. Most farmers and hatchery managers have been trained to be efficient, and this means spawning the fewest number of fish that will enable them to meet production goals. Farmers in particular want to be efficient so that they spend less money raising their crop. Furthermore, one way hatchery managers at public hatcheries are evaluated is cost per fingerling. But this philosophy is at odds with the control of inbreeding and genetic drift.

This problem is especially acute in aquaculture, because the fecundity of some species of fish is so great that it is often possible to spawn one or two females and males and produce the number of fingerlings that are needed for grow-out. But the ability to spawn relatively few fish must be moderated if inbreeding and genetic drift are to be controlled.

The best way to manage a population is to determine what N_e is needed to prevent inbreeding- and genetic drift-related problems under the time frame that has been chosen and then spawn the number of brood fish that will enable the farmer to produce that N_e. The farmer should spawn many more fish than he normally would and simply keep a small and equal random sample of eggs or fingerlings from each spawn. He can sell the surplus eggs or fish to other farmers. If he cannot sell the surplus fish, they should be discarded. This might seem wasteful, but this practice improves the genetics of the population.

Spawn a More Equal Sex Ratio

An easy way to increase N_e and thus reduce the rate of inbreeding and genetic drift is to spawn a more equal sex ratio. Most farmers and hatchery managers use skewed sex ratios when they spawn their fish. This is done because one male can usually be used to fertilize eggs from several females. This enables aquaculturists to use and maintain fewer males, which lowers production costs. This may be a great idea in terms of efficiency, but it is a bad idea if you want to manage and control the genetic quality of the population. When a skewed sex ratio is used, the rarer sex has a disproportionate influence on the size of N_e.

The impact that a skewed sex ratio has on inbreeding can be seen in the following formula:

$$F = \frac{1}{8(number\ of\ females)} + \frac{1}{8(number\ of\ males)}$$

Where: number of males and number of females are the numbers that produce viable offspring. The effect that a skewed sex ratio has on inbreeding is more pronounced when the total breeding population is small. For example, if a farmer spawns 100 females and 10 males he produces the following amount of inbreeding:

$$F = \frac{1}{8(100)} + \frac{1}{8(10)}$$

F=0.01375

If he moderates the sex ratio and uses 100 females and 40 males the inbreeding becomes:

$$F = \frac{1}{8(100)} + \frac{1}{8(10)}$$

F=0.004375

The addition of 30 males reduced inbreeding per generation from 1.38% to 0.44%.

The impact that skewed sex ratios have on inbreeding even means that smaller populations can produce less inbreeding. For example, a population with 50 males and 50 females produces less inbreeding than one with 20 males and 200 females (0.5% vs 0.69%) even though it is less than half as large.

The inbreeding that is produced by various combinations of males and females is shown. The inbreeding values in that table show that if the number of one sex is held constant, increasing the other sex produces diminishing returns; i.e., the reduction in inbreeding becomes less and less despite substantial increases in the commoner sex. For example, if a farmer spawns 10 males and 100 females, F = 1.38%; if he spawns 10 males and 250 females, F = 1.30%.

Fertilization Techniques

A final approach to managing N_e is to alter certain hatchery practices if gametes are stripped. When eggs and sperm are stripped, many farmers and hatchery managers pool eggs from several females and then fertilize the eggs either by using pooled milt or by adding milt from several males in a sequential manner. Both techniques will lower N_e from what you think it is. The reasons are: One, if milt is pooled and added, there is competition among the sperm, due to a difference in potency. Some males produce more vigorous sperm, and these sperm cells will fertilize a disproportionate number of eggs. In some cases, most of the eggs will be fertilized by sperm from a single male. Second, if milt is added in a sequential manner, the sperm from the first male will fertilize most of the eggs, while that from the last few males will fertilize comparatively few eggs.

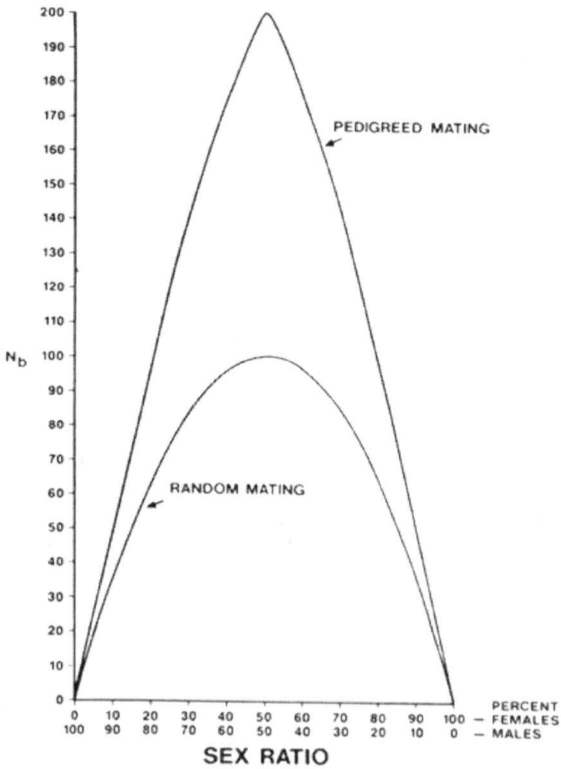

Figure: Effective breeding efficiency (N_b) over all possible sex ratios for both random mating and pedigreed mating. Effective breeding efficiency is a measure of the efficiency of the sex ratio or mating system in maximising N_e within the constraints of a fixed population size (N).

The solutions to both problems are quite simple. Fertilize eggs from each female with sperm from a single male. If you want to have sperm from several males fertilize eggs from each female, subdivide each egg mass and fertilize each lot separately. If this is done, N_e can be accurately determined, because you know which brood fish produced offspring.

Unless a farmer is going to conduct a selective breeding programme or use inbreeding to improve the results of selection or crossbreeding, a population should be managed genetically to prevent unwanted inbreeding from causing inbreeding depression and to prevent genetic drift from robbing the population of alleles and genetic variance. If fish can be marked, inbreeding depression can be prevented by creating pedigrees and by preventing consanguineous matings or by preventing matings between relatives more closely related than second cousins.

Marking fish and preventing consanguineous matings will not prevent genetic drift. Managing a population to minimise the effects of genetic drift can be accomplished only by managing N_e.

When fish are not marked (which will be the case for most hatchery populations), the only way to prevent unwanted inbreeding from accumulating and to prevent the ravages of genetic drift is to manage the population's N_e. When a farmer is not using a breeding programme to improve a population, managing a population's N_e is probably the most important aspect of brood stock management. A population's N_e is one of the most important pieces of information about the population, because N_e is inversely related to both inbreeding and genetic drift. Consequently, managing N_e is a key aspect of fish husbandry.

There is no universal N_e that can be used to manage every population. It must be customized for each population. This chapter outlined the techniques and methods that must be used to determine the N_e that is needed. That number can be determined by answering a series of questions: One, what level of inbreeding will cause problems? Two, what is the frequency of the rare alleles that a farmer wants to save, and what guarantee does he want that the alleles have been saved? Three, how many generations does the farmer want to incorporate into the work plan before that level of inbreeding has been reached and when the guarantee will end?

The levels of inbreeding that cause problems in hatchery populations are not known, so appropriate levels must be determined by a "guesstimate." Two values were proposed: 5% and 10%. The value chosen depends on how important the population is and what the goals are. Populations that are being cultured for food can use either 5% or 10%. Populations that are being cultured for stocking programmes must use 5%; if possible, lower levels should be used for these populations.

The frequency of the rare alleles that should be saved depends on the population. If the fish are being farmed, the frequency should be 0.05 for most farmers who wish to manage their populations genetically, and 0.01 for those who have the ability and desire to conserve as much genetic variance as possible. If the population is being cultured for stocking programmes the frequency should be no greater than 0.01; if possible, the frequency should be 0.005-0.001. These frequencies are not absolute but are presented as guideline values. When managing a population's N_e, the major goal is to maintain N_e at a constant level every generation. If N_e drops below the desired value for a single generation, the genetic goals cannot be achieved. Maintaining N_e at the desired level generation after generation may be the most difficult aspect of brood stock management, because N_e can decline for a variety of reasons. Sudden and drastic decreases in N_e are called bottlenecks, and they can cause permanent and irreversible genetic damage.

Finally, there are a number of management techniques that can be used to increase N_e or that can be used to produce the same level of inbreeding but with a smaller number of brood fish: pedigreed mating; stretching generations; bringing the sex ratio closer to a 1:1 ratio; and altering some fertilization techniques when gametes are stripped.

Results and Discussion

The amplification products did not demonstrate any sequence heterogeneity supporting the existence of two msGHs, as has been found in *Tilapia nilotica* (Ber and Daniel, 1993). The msGH cDNA contains an open reading frame of 615 nucleotides starting at the first ATG codon located at 109 nucleotides from the 5' terminus of the mRNA and ending with a TGA stop codon, encoding a preprotein of 204 amino acids residues as shown above. The msGH mRNA 3' untranslated region is 204 nucleotides long and it contains two polyadenylation consensus sequences AATAAA.

This is the first time an atherinid growth hormone-encoding cDNA has been reported and the deduced amino acid sequence obtained.

In addition, no data is available regarding the amino acid sequence of the authentic msGH polypeptide. Since the first 17 amino acid residues from the N terminus are highly hydrophobic (70% of the amino acids residues of this region are non-polar) and also have a high degree of homology to the signal peptide of other fish GHs, it is assumed that in the msGH this region probably represents the signal peptide of the pre-GH which is cleaved upon hormone secretion.

The hormone exhibits typical GH features, such as four cysteine residues, capable of forming two disulphide bonds that are assumed to contribute to the tertiary structure of the hormone molecule, a single tryptophan residue and stretches of amino acids highly conserved in all known GHs. There is only one Asn-Xaa-Thr motif in the msfGH aa sequence (Asn^{201}) which is a potential site for N-linked glycosylation. The mature form of msGH deduced from the msGH cDNA sequence contains 187 amino acids, starting with a glutamine. The amino acid sequence, when compared to those of mature GHs from other acanthopterygian fishes, showed higher levels of homology to fish GHs representing the order Perciformes: with a similarity of up to 92% (data not shown).

The consensus phylogenetic tree obtained showed the major phylogenetic grouping of euteleosts supporting the present classification (Nelson, 1994). Members of the Ostariophysi (*I. Punctatus*, *H. nobilis* and *C. carpio*), Protacanthopterygii (*S. salar* and *O. mykiss*) and Paracanthopterygii (*G. morhua*) Superorders grouped according to the current fish classification. Within Acanthopterygii only Pleuronectiformes (*S. senegalensis*, *P. olivaceus* and *V. variegatus*) grouped as a monophyletic lineage. The remaining orders Atheriniformes (*O. argentinensis*), Scorpaeniformes (*S. schlegeli*), Tetraodontiformes (*F. rubripes*) and Perciformes (*S. aurata*, *T. thynnus*, *T. nilotica*, *L. calcarifer*, *S. ocellatus* and *M. saxatilis*) did not revel any statistically supported relationship, in spite of the high similarity observed between Perciformes and Atheriniformes GHs at the amino acid sequence level.

Since the atherinomorph classification was created by Rosen (1964), several hypotheses about sister-group and family relationship have been suggested based only on morphological characters. Gosline (1971), pointed out the affinity of this group with members of the former perciform Suborder Mugiloidei, which is currently in the Mugilomorpha Series, Mugiliformes Order (Stiassny, 1990). This author proposed that atherinomorphs and mugiloids are sister taxa based in four atherinomorph/mugiloid synapomorphies, even though he made some reservations regarding mugiloids having the percomorph type pelvic girdle which represents a character conflict, suggesting the need for new approaches to solve this question (Stiassny, 1993). Parenti (1993) suggested that atherinomorphs are the sister group of some or all paracanthopterygian fishes.

The same author stated that the relationship between atherinomorphs, percomorphs and paracanthopterygians is not resolved because the data are incomplete, arguing that atherinomorphs/percomorphs sister group relationship is barely supported. However, atheriniform fishes were considered to be composed of two monophyletic groups: Atherinopsidae, the New World silversides, and Atherinidae, the Old World silversides both of which have percoids and mugilids as sister-groups (Saeed *et al.*, 1994; Dyer and Chernoff, 1996). Recently, Wiley *et al.* (2000) in a comprehensive study about the interrelationships of acanthomorph fishes using molecular and morphological data, showed evidence corroborating the Stiassny's (1993) hypothesis that mullets are related to atherinomorphs. Even though there are many hypotheses for atherinomorphs relationship, our results only allow us to refute the hypothesis that atherinids are related to paracanthopterygians as previously suggested (Parenti, 1993). Within Acanthopterygii we observed a lack of solid synapomorphies among most of the Orders analysed, indicating a complex assemblage of fishes in which the phylogenetic tree remains indeterminable.

However, based on the agreement of the molecular data obtained in the present work with the major fish groups currently accepted phylogenies and the more recent approaches using molecular and morphological data, we are sure that nuclear genes such as the growth hormone gene can be useful to solve the question about sister-group relationships in fish. In addition, we would have a better view of this situation if new information at the molecular level was gathered, mainly coming from species representing fish groups that have no such data currently available such as Mugilomorpha (Mugiliformes Order), as well as the remaining Atherinomorpha Orders (Orders Beloniformes and Cyprinodontiformes Orders) and Percomorpha.

Chapter 5

Cell Biology of Oxytocin and Vasopressin Cells in Fishes

The neurohypophysial peptide oxytocin (OT) and OT-like hormones facilitate reproduction in all vertebrates at several levels. The major site of OT gene expression is the magnocellular neurons of the hypothalamic paraventricular and supraoptic nuclei. In response to a variety of stimuli such as suckling, parturition, or certain kinds of stress, the processed OT peptide is released from the posterior pituitary into the systemic circulation. Such stimuli also lead to an intranuclear release of OT. Moreover, oxytocinergic neurons display widespread projections throughout the central nervous system. However, OT is also synthesized in peripheral tissues, e.g., uterus, placenta, amnion, corpus luteum, testis, and heart. The OT receptor is a typical class I G protein-coupled receptor that is primarily coupled via G_q proteins to phospholipase C-β.

The high-affinity receptor state requires both Mg^{2+} and cholesterol, which probably function as allosteric modulators. The agonist-binding region of the receptor has been characterised by mutagenesis and molecular modelling and is different from the antagonist binding site. The function and physiological regulation of the OT system is strongly steroid dependent. However, this is, unexpectedly, only partially reflected by the promoter sequences in the OT receptor gene. The classical actions of OT are stimulation of uterine smooth muscle contraction during labour and milk ejection during lactation. While the essential role of OT for the milk let-down reflex has been confirmed in OT-deficient mice, OT's role in parturition is obviously more complex. Before the onset of labour, uterine sensitivity to OT markedly increases concomitant with a strong upregulation of OT receptors in the myometrium and, to a lesser extent, in the decidua where OT stimulates the release of $PGF_{2\alpha}$. Experiments with transgenic mice suggest that OT acts as a luteotrophic hormone opposing the luteolytic action of

$PGF_{2\alpha}$. Thus, to initiate labour, it might be essential to generate sufficient $PGF_{2\alpha}$ to overcome the luteotrophic action of OT in late gestation. OT also plays an important role in many other reproduction-related functions, such as control of the estrous cycle length, follicle luteinization in the ovary, and ovarian steroidogenesis. In the male, OT is a potent stimulator of spontaneous erections in rats and is involved in ejaculation. OT receptors have also been identified in other tissues, including the kidney, heart, thymus, pancreas, and adipocytes. For example, in the rat, OT is a cardiovascular hormone acting in concert with atrial natriuretic peptide to induce natriuresis and kaliuresis. The central actions of OT range from the modulation of the neuroendocrine reflexes to the establishment of complex social and bonding behaviours related to the reproduction and care of the offspring. OT exerts potent antistress effects that may facilitate pair bonds. Overall, the regulation by gonadal and adrenal steroids is one of the most remarkable features of the OT system and is, unfortunately, the least understood. One has to conclude that the physiological regulation of the OT system will remain puzzling as long as the molecular mechanisms of genomic and nongenomic actions of steroids have not been clarified.

The neurohypophysial hormone oxytocin (OT) was the first peptide hormone to have its structure determined and the first to be chemically synthesized in biologically active form. It is named after the "quick birth" which it causes due to its uterotonic activity. OT was also found to be responsible for the milk-ejecting activity of the posterior pituitary gland. The structure of the OT gene was elucidated in 1984, and the sequence of the OT receptor was reported in 1992.

OT is a very abundant neuropeptide. This became obvious in a study where the most prevalent hypothalamic-specific mRNAs were analysed. OT was found to be the most abundant of 43 transcripts identified. Today, we recognise that OT exerts a wide spectrum of central and peripheral effects. The actions of OT range from the modulation of neuroendocrine reflexes to the establishment of complex social and bonding behaviours related to the reproduction and care of the offspring. Overall, the cyclic nonapeptide OT and its structurally related peptides facilitate the reproduction in all vertebrates at several levels.

In the following two sections, we describe the structural features of the OT receptor system on the molecular level. OT has long been considered to be restricted to stimulation of uterine contractions during labour and milk ejection during lactation. These classical functions of the OT system are treated in the first parts of section IV. The fact that OT is found in equivalent concentrations in the neurohypophysis and plasma of both sexes suggests that OT has further physiological functions. The expression of OT and its receptor has now been identified in a variety of peripheral tissues. For example, evidence was provided that OT acts in concert with atrial natriuretic peptide (ANP) in the control of body fluid and in cardiovascular homeostasis in the rat. It is important to note that most of our knowledge about functions of OT derives from studies with rats. However, localisation and expression patterns of OT receptors show marked species differences, suggesting that some of the described OT activities may be species specific. Over the past decade, particularly the central actions of OT have been intensively studied revealing a profound regulation by steroids. The regulation by steroids is in fact a common theme throughout the review and is probably the most remarkable feature of the OT receptor system. Finally, in the last section, we summarise the contributions that have been reported on behavioural effects mediated or modulated by OT. Ironically, the classical "oxytocic" function of OT is again open for discussion due to the results with OT-deficient mice.

Oxytocin and Oxytocin-like Peptides

Evolutionary Aspects A

All neurohypophysial hormones are nonapeptides with a disulfide bridge between Cys residues 1 and 6. This results in a peptide constituted of a six-amino acid cyclic part and a COOH-terminal α-amidated three-residue tail. Based on the amino acid at position 8, these peptides are classified into vasopressin and OT families: the vasopressin family contains a basic amino acid (Lys, Arg), and the OT family contains a neutral amino acid at this position. Isoleucine in position 3 is essential for stimulating OT receptors and Arg or Lys in position 8 for acting on vasopressin receptors. The difference in the polarity of these amino acid residues is believed to enable the vasopressin and OT peptides to interact with the respective receptors.

Virtually all vertebrate species possess an OT-like and a vasopressin-like peptide. Bony fishes (Osteichthyes), predecessors of the land vertebrates, possess isotocin and vasotocin. Thus two evolutionary molecular lineages have been proposed: an isotocin-mesotocin-OT line, associated with reproductive functions, and a vasotocin-vasopressin line involved in water homeostasis. Because vasotocin has been found in the most primitive cyclostomes, the OT and vasopressin genes may have arose by duplication of a common ancestral gene after the radiation of cyclostomes. Based on calculations from the nucleotide level of OT and vasopressin gene, the ancestral gene encoding the precursor protein should be more than 500 million years old. The exceptional structural stability of the nonapeptides during evolution suggests a strong selective pressure, e.g., by coevolution with the corresponding receptors and/or with specific processing enzymes. A conspicuous diversity of OT-like peptides is found in cartilaginous fishes (Chondrichthyes). These marine fishes use urea rather than salts for osmoregulation. It was hypothesized that the OT-like hormones gained their high diversity in Chondrichthyes as they have been relieved from the control of ionic homeostasis. Notably, OT, the typical hormone of placental mammals, has been identified in the Pacific ratfish, a Chondrichthyes species.

Mesotocin is the OT-like hormone found in most terrestrial vertebrates from lungfishes to marsupials, which includes all nonmammalian tetrapods (amphibians, reptiles, and birds). Only two South American marsupials express OT exclusively, whereas all other marsupials have mesotocin. In the Northern brown bandicoot (*Isoodon macrourus*) and the North American opossum (*Didelphis virginiana*), OT is present together with mesotocin. Overall, mesotocin has the largest distribution in vertebrates after vasotocin found in all nonmammalian vertebrates and isotocin identified in bony fishes. Despite this invariability, no clear physiological role has been ascribed to this peptide so far. It is unknown whether the marsupial species that are endowed with both OT and mesotocin have two distinct receptors. The earthworm *Eisenia foetida* is the most primitive species from which an OT-related peptide (annetocin) has been isolated. Injection of annetocin in the earthworm or in leechs results in induction of egg-laying behaviour.

Gene Structure B

In all species, OT and vasopressin genes are on the same chromosomal locus but are transcribed in opposite directions . The intergenic distance between these genes range from 3 to 12 kb in mouse, human, and rat. This type of genomic arrangement could result from the duplication of a common ancestral gene, which was followed by inversion of one of the genes. The human gene for OT-neurophysin I encoding the OT prepropeptide is mapped to chromosome 20p13 and consists of three exons: the first exon encodes a translocator signal, the nonapeptide hormone, the tripeptide processing signal

(GKR), and the first nine residues of neurophysin; the second exon encodes the central part of neurophysin (residues 10–76); and the third exon encodes the COOH-terminal region of neurophysin (residues 77–93/95).

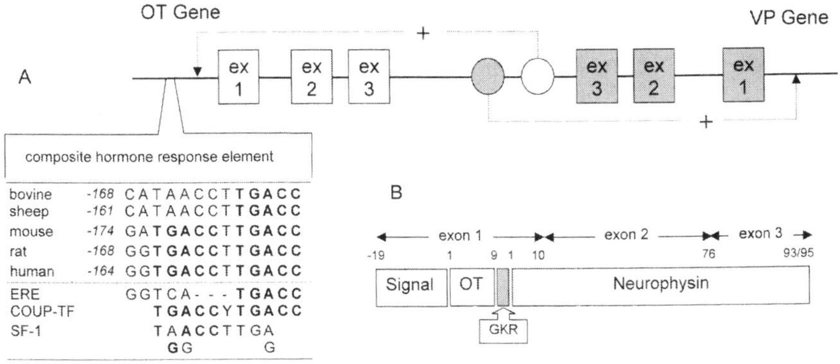

Figure: Organisation of the oxytocin (OT) and vasopressin (VP) gene structure including schematic depiction of the putative cell-specific enhancers (open circle, enhancer of OT gene; shaded circle, enhancer of VP gene). *A*: details of the approximately "160-bp region (composite hormone response element) of the upstream OT gene promoter conserved across five species including the sequences of the response elements estrogen response element (ERE), chicken ovalbumin upstream promoter transcription factor I (COUP-TF), and steroidogenic factor-1 (SF-1) are indicated. *B*: domain organisation of preprooxytocin including the processing sites. The precursor is split into the indicated fragments by enzymatic cleavages, one involving a glycyl-lysyl-arginine (GKR) sequence and leaving a carboxamide group at the COOH-terminal end of OT. Signal, signal peptide.

The high homology of the OT-like precursor polypeptides is well documented in the sequence of preproannetocin from *Eisenia foetida*, a primitive invertebrate. It consists of a signal peptide, annetocin (flanked by a Gly COOH-terminal amidation signal and a Lys-Arg dibasic endoproteolytic sequence), and a neurophysin domain. Notably, 14 cysteine residues that play a crucial role in constructing the correct tertiary structure of a neurophysin are completely conserved in the *Eisenia* neurophysin domain.

The OT prepropeptide is subject to cleavage and other modifications as it is transported down the axon to terminals located in the posterior pituitary. The mature peptide products, OT and its carrier molecule neurophysin, are stored in the axon terminals until neural inputs elicit their release. The main function of neurophysin, a small (93–95 residues) disulfide-rich protein, appears to be related to the proper targeting, packaging, and storage of OT within the granula before release into the bloodstream. OT is found in high concentrations (>0.1 M) in the neurosecretory granules of the posterior pituitary complexed in a 1:1 ratio with neurophysin. In such complexes, OT-neurophysin dimmers are the basic functional units as suggested by the crystal structure of the neurophysin-OT complex. Cys-1 and Tyr-2 in the OT molecule are the principal neurophysin binding residues. In particular, the protonated α-amino group (Cys-1) in OT forms an essential contact site to neurophysin via electrostatic and multiple hydrogen bonding interactions. Due to its dependence on amino group protonation (pK_a ~6.4), the binding strength between OT and neurophysin is much higher in an acidic compartment like the neurosecretory granules (pH ~5.5). Conversely, the dissociation of the complex is facilitated as the complex is released from the neurosecretory granules and enters the plasma (pH 7.4).

Gene Regulation C

Due to the lack of an appropriate cell culture system, the regulation of OT gene expression was studied in heterologous systems and in transgenic mice. The expression patterns found with bovine OT transgenes in mice suggested very complex mechanisms for the cell type-specific expression of OT genes. A bovine OT transgene consisting of the OT structural gene flanked by 600 bp of upstream and 1,900 bp of downstream sequences contained sufficient information to direct expression to murine oxytocinergic magnocellular neurons, within which it was subject to physiological regulation. However, enlargement of OT transgene constructs by addition of 700 bp of contiguous downstream sequences repressed the hypothalamic expression. Analysis of various gene constructs in transgenic mice led to the proposal that cell-specific enhancers for OT and vasopressin gene expression are not located on the 52 -upstream regions of these genes, but are present in the intergenic region 0.5–3 kb downstream of the vasopressin gene. So, constructs containing genomic DNA from 0.5 to 9 kb 52 -upstream of the OT and vasopressin genes but with no endogeneous 32 -downstream sequences did not show significant expression in the hypothalamic magnocellular neurons.

The OT mRNA in the rat shows an increase in poly (A) tail length in response to the activation of the hypothalamoneurohypophysial system, e.g., during pregnancy, lactation, and dehydration. This could augment mRNA stability and may be an additional level of OT gene control. Hexanucleotide AGGTCA motifs and variations thereof are present in the proximal 52 -flanking region of cloned OT genes. This motif is part of binding sites for all members of the nuclear receptor superfamily, except the glucocorticoid, mineralocorticoid, progesterone, and androgen receptor. Various combinations of this motif exist, ranging from single hexanucleotides, direct or inverted repeats with spacing varying from one to at least six nucleotides. Thus potentially several members of the nuclear receptor family including many orphan receptors could interact with the OT gene and regulate its expression. The human and rat OT promoters could be stimulated by the ligand-activated estrogen receptors ERα and ERβ, the thyroid hormone receptor THRα, and the retinoic acid receptors RARα and RARβ in a variety of cells. However, it is important to note that these results were obtained from cotransfection experiments in cell lines, i.e., under nonphysiological circumstances.

A highly conserved DNA element exists at -160 nucleotides upstream from the transcriptional initiation site. Deleting the region between 172 and 148 resulted in complete loss of thyroid hormone responsiveness and most of the responsiveness to estrogen and retinoic acid. This special "composite" hormone response element is composed of three TGACC motifs. Two of them form an inverted repeat with a spacing of three nucleotides that differs in one nucleotide from the palindromic, canonical estrogen response element (ERE). The rat and human OT promoter shows a good homology with the classic palindromic ERE. Accordingly, in a heterologous transfection system, the rat and human but not the bovine OT gene promoter could be stimulated by estradiol. In the appropriate cellular context, the strongest activators were ERα and ERβ. The composite hormone response element was suggested to synergise with proximal elements for the estrogen responsiveness and was found to be essential for the positive regulation by retinoic acid. However, estrogen receptor expression was not detected in oxytocinergic cells of the rat hypothalamus. Thus direct estrogen-dependent activation may not regulate the OT gene expression in magnocellular neurons in vivo. An estrogen responsiveness was reported for a parvocellular OT-expressing cell group that contains ERβ. Although THRs have been localised in supraoptic nucleus (SON) and paraventricular nucleus (PVN), only a small influence of thyroid hormones on OT gene expression was found in rats in vivo.

The rat uterus displays a marked upregulation of OT gene expression before delivery. The main site of steroid-induced uterine OT gene expression was the endometrial epithelium. A strong increase in OT mRNA (150 fold) preceded the increase in uterine OT binding sites that occurs very shortly before the onset of labour. The estrogen-induced rise in uterine OT mRNA was probably mediated via the common hormone response element in the OT gene promoter. The palindromic structure at the composite hormone response element at approximately –160 bp was identified as necessary and sufficient for estrogen induction of the OT gene promoter. However, the high level of uterine OT mRNA at term was not achieved by any of the steroid treatment regimens tested so far.

Several investigations have focused on the role of nuclear orphan receptors in the regulation of the OT gene. Unlike in the hypothalamus, the bovine OT gene could be tissue-specifically expressed in the gonads by a minimal functional promoter contained within 600 bp of the transcription start site. As mentioned above, the bovine OT gene is unresponsive to estradiol. Nuclear orphan receptors of corpus luteum granule cells have been identified to interact with the common hormone response element: chicken ovalbumin upstream promoter transcription factor I (COUP-TFI) and steroidogenic factor-1 (SF-1). The levels of these factors could be responsible for the regulation of the endogeneous OT gene in this tissue. SF-1 is a factor with constitutive activating properties on the OT gene. The orphan COUP-TFI repressed the activation of the rat OT gene induced by retinoic acid, thyroid hormone, and estrogens through competitive binding to the composite hormone response element. Another orphan receptor identified in the hypothalamus, testis receptor 4 (TR4), interacts, unlike all other nuclear receptors, at a region further downstream (–112/"77 bp) of the common response element in the OT gene. The 52 -flanked region of the rat OT gene also contains binding sites for class III POU homeodomain proteins. Brn-2, a member of this family, is involved in the regulation of OT genes in magnocellular neurons. Brn-2 null mice lack magnocellular vasopressin- and OT-expressing neurons in SON and PVN. Brn-2 is required for a specific step in the developmental fate of magnocellular neurons. However, POU class III proteins did not display a significant regulatory activity on the OT gene in heterologous expression systems. Taken together, OT gene regulation in vivo appears to be governed by multiple enhancers and repressors interacting in a complex yet ill-defined fashion.

Oxytocin Receptors

Gene Structure and Regulation A.

Kimura et al. first isolated and identified a cDNA encoding the human OT receptor using an expression cloning strategy. The encoded receptor is a 389-amino acid polypeptide with 7 transmembrane domains and belongs to the class I G protein-coupled receptor (GPCR) family. To date, the OT receptor encoding sequences from pig, rat, sheep, bovine, mouse, and rhesus monkey have also been identified.

The human OT receptor mRNAs were found to be of two sizes, 3.6 kb in breast and 4.4 kb in ovary, endometrium, and myometrium. The OT receptor gene is present in single copy in the human genome and was mapped to the gene locus 3p25-3p26.2. The gene spans 17 kb and contains 3 introns and 4 exons. Exons 1 and 2 correspond to the 52 -prime noncoding region. Exons 3 and 4 encode the amino acids of the OT receptor. Intron 3, which is the largest at 12 kb, separates the coding region immediately after the putative transmembrane domain 6. Exon 4 contains the sequence encoding the seventh transmembrane domain, the COOH terminus, and the entire 32 noncoding region, inluding the polyadenylation signals. Although many GPCRs have an intronless gene structure, the genes for some other members of the GPCR family including the human vasopressin V2 receptor contain an intron at the same location after transmembrane domain 6. The transcription start sites lie 618 and 621 bp upstream of the initiation codon as demonstrated by primer extension analysis. Nearby, a TATA-like motif and a potential SP-1 binding site is found in the human OT receptor gene. The 52 -flanking region also contains invert GATA-1 motifs, one c-Myb binding site, one AP-2 site, two AP-1 sites, but no complete ERE. Instead, there were two half-palindromic 52 -GGTCA-32 motifs and one half-palindromic 52 -TGACC-32 motif of ERE. Moreover, there were two nucleofactor interleukin-6 (NF IL-6) binding consensus sequences and two binding site sequences for an acute phase reactant-responsive element at the 52 -flanking region.

In the OT receptor gene of the mouse, the promoter region lacks an apparent TATA box but contains multiple putative interleukin-response elements, several half-palindromic motifs, and a classical ERE.

In case of the rat OT receptor gene expression at parturition, three transcripts (2.9, 4.8, and 6.7 kb) were identified that differ in the length of their 32 -untranslated regions. The promoter region of the rat OT receptor gene also contains multiple putative interleukin-response elements, NF IL-6, and acute-phase response elements (APRE). Further sequence analysis of 4 kb of the 52 -flanking DNA of the rat OT receptor gene revealed the presence of a cAMP response element (CRE) as well as several other potential regulatory elements, including AP-1, AP-2, AP-3, AP-4 sites, an ERE, and a half-steroid response element. A palindromic ERE was identified -4 kb 52 of the translational start site. An OT receptor reporter construct of this promoter in the human cancer breast cell line MCF-7 demonstrated pronounced induction by both forskolin and phorbol ester, but contrary to in vivo findings, only a weak transcriptional response to estradiol.

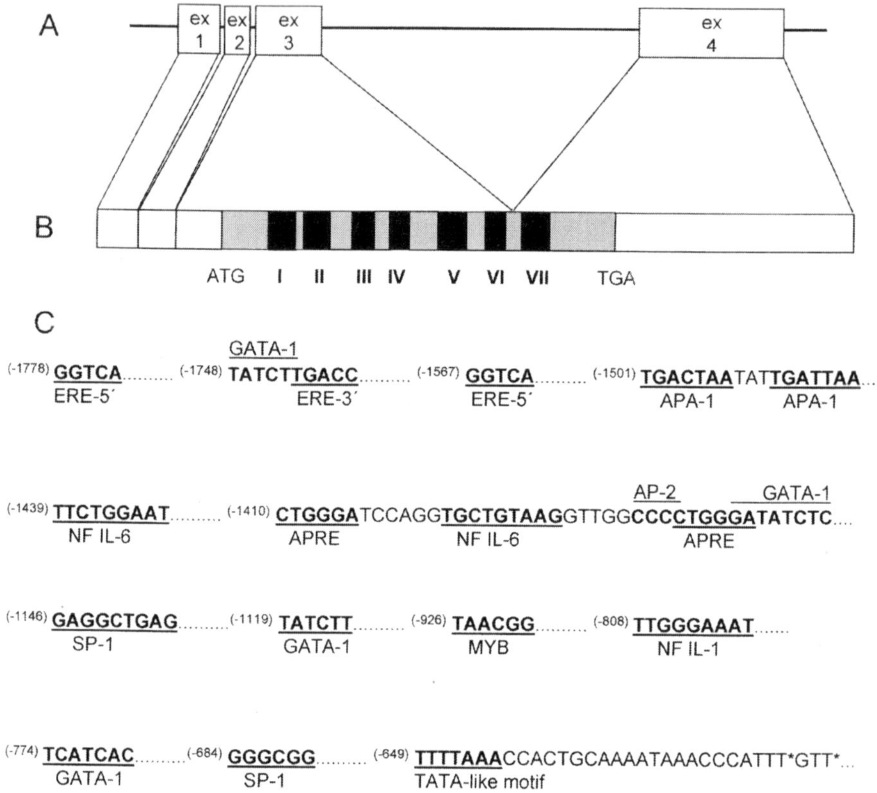

Figure: Organisation of the human OT receptor gene including the localisation of consensus sequences for transcription factors. The human OT receptor gene consists of four exons. Exons 3 and 4 encode the amino acid sequence for the OT receptor. The start (ATG) and stop (TGA) codons of the receptor cDNA are indicated. The DNA sequences encoding for transmembrane regions I–VII are indicated by black areas.

Constructs of the CRE and half-steroid response elements from the promoter of the rat OT receptor gene function as active enhancers. This suggests a potential role for protein kinase A and C pathways in OT receptor gene regulation. The protein kinase A pathway, for example, may be activated when forskolin treatment promotes upregulation of OT receptors in cultured rabbit amnion cells. Protein kinase C may act to increase fos/jun activity at AP-1 sites in response to phorbol ester treatment. In a mammary tumor cell line (Hs578T) that expresses inducible, endogenous OT receptors, a DNA region containing an ets family target sequence, and a CRE/AP-1-like motif was required for both basal and serum-induced OT receptor gene expression. The ets factor GABPα/β slightly induced OT receptor gene expression in this human breast cell line. The gene expression was markedly potentiated following cotransfection with c-*fos*/c-*jun*.

APREs are typically found in genes for acute-phase proteins such as α_2-macroglobulin or T kininogen, which are induced by infection or inflammation. The presence of these elements in the promoter region of the human and rat OT receptor gene suggests that the acute induction of OT receptor expression could be a phenomenon similar to the induction of acute-phase reponse genes. The decidua has "macrophage-like" properties and functions. Possibly, inflammatory cytokines are able to induce labour and, thereby, take usage of the transcriptional activation of the OT receptor gene.

However, the lack of classical EREs in a promoter does not exclude a potential direct effect of estrogens on gene expression, since the present half-palindromic ERE motifs can also act synergistically to mediate estrogen activation as shown in the ovalbumin gene. Gonadal steroids have an important influence on the uterine OT receptor mRNA accumulation in vivo. Estrogens administered to ovariectomised rats increased OT receptor binding sites and increased OT receptor mRNA accumulation severalfold. Although progesterone leads to a marked decline of OT receptor binding sites, the mRNA levels of OT receptor were nearly unchanged. This and several other findings imply the involvement of nongenomic effects of progesterone.

The OT receptor gene is differentially expressed in various tissues. In uterus or hypothalamus, the OT receptor regulation correlates with the pattern of sex steroids, in particular estradiol. As shown with knock-out mice, ERα is not necessary for basal OT receptor synthesis but is absolutely necessary for the induction of OT receptor binding in the brain by estrogen. However, it is unclear whether OT receptor gene transcription is predominantly regulated by estrogen. The continuous presence of receptors in certain brain regions after gonadectomy suggests the existence of alternate mechanisms of regulation. In this context, a study of the tammar wallaby, an Australian marsupial, is interesting. This species has a twin uterus attached to a double cervix so that each uterus forms an independent environment. During pregnancy, only one uterus becomes gravid, the other remains empty and can thus be regarded as a natural control for uterine changes occurring during pregnancy. It was shown that mesotocin receptor concentrations and the responsiveness to mesotocin differed between the gravid and nongravid myometrium during pregnancy. This indicates that the stimulating agent for the mesotocin receptor is unlikely a circulating factor but rather a local factor, possibly of fetal or placental origin. Another well-studied system is the OT receptor gene expression in bovine endometrial cells. In vivo, bovine endometrial OT receptors are upregulated in a cycle-dependent fashion. This regulation appears to be completely at the transcriptional level. Even if the receptors are downregulated in vivo, they show upregulation when explanted and cultured in vitro. This indicates that the OT receptor regulation is partly due to gene suppression in vivo. Despite the presence of steroid receptors in bovine endometrial cells, the level of OT receptor mRNA could neither be affected by progesterone or estradiol nor by a progesterone withdrawal protocol. The only factor that affected the OT receptor mRNA level was interferon-[3]. As in vivo, this cytokine suppressed the OT receptor mRNA production. Nuclear protein binding and transfection experiments suggested that constitutive upregulation is a feature of the OT receptor promoter. So, specific gene suppression is likely to play an important role for physiological control of the OT receptor expression. Of interest, a genomic element within the third intron of the human OT receptor gene was found to be associated with transcriptional gene suppression. The intronic region was hypermethylated in nonexpressing tissues, but relatively hypomethylated in the myometrium of the cycle and at term, when the OT receptor gene is upregulated. Taken together, it was concluded that sex steroids have an indirect effect on both the OT and OT receptor genes, possibly involving intermediate transcription factors or cofactors. The transcriptional regulation of OT receptor shows species-specific differences. The brain OT receptor varies across species in its distribution as well as in its regional regulation by gonadal steroids. For the OT receptor as for many other genes, the DNA sequences located in the 52 -flanking region upstream from the coding region are primarily responsible for conferring tissue-specific expression. Transgenic mice carrying 5 kb of the 52 -flanking region of the prairie vole OT receptor gene showed the typical expression pattern of prairie vole OT receptors in mice. However, the regulatory elements that confer the specific expression patterns for the OT receptor gene in vivo are yet unknown.

One of the fundamental questions concerns the possible existence of OT receptor subtypes. Such subtypes have been suggested to be present, e.g., in the rat uterus, kidney, or brain, to explain differential pharmacological profiles or immunoreactivity patterns. Application of polymerase chain reaction methods and Southern analysis in several tissues known to possess OT binding activity failed to identify a gene encoding a further OT receptor subtype. However, the applied techniques only screen for genes with high homology to the uterine-type OT receptor, and therefore, a putative further OT receptor with low homology to the uterine-type OT receptor would have been kept undetected.

Receptor Structure B

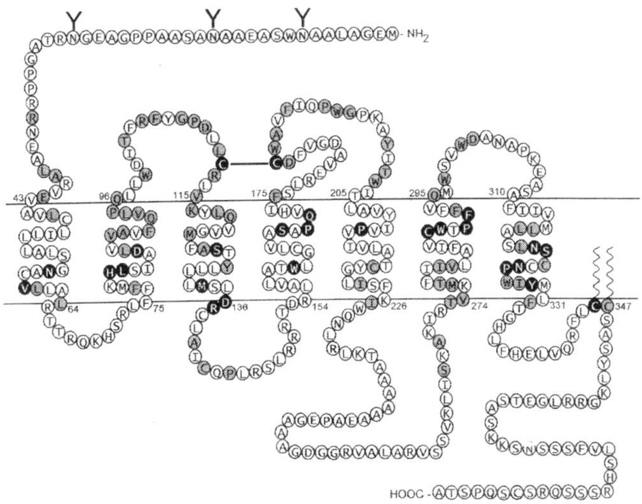

Figure: Schematic structure of the human OT receptor with amino acid residues shown in one-letter code. The residues are marked in the same manner as in Fig. 3, i.e., residues conservative within the OT/vasopressin receptor subfamily are outlined in gray, and residues conservative for the whole G protein-coupled receptor superfamily are outlined in black. The putative *N*-glycosylation ("Y") and palmitoylation (at C346/C347) sites are marked.

The OT receptor is a typical member of the rhodopsin-type (class I) GPCR family. The seven transmembrane α-helices are most highly conserved among the GPCR family members. Conserved residues among the GPCRs may be involved in a common mechanism for activation and signal transduction to the G protein. On the basis of studies with model GPCRs, it is assumed that the switching from the inactive to the active conformation is associated with a change in the relative orientation of transmembrane domains 3 and 6, which then unmasks G protein binding sites. In the class I GPCR family, an Asp in transmembrane domain 2 and a tripeptide (E/D RY) at the interface of transmembrane 2 and the first intracellular loop are believed to be important for receptor activation. With respect to Asp-85, this was confirmed for the human OT receptor. When Asp-85 is exchanged by the residues Asn, Gln, or Ala, agonist binding and signal transduction of the receptor becomes impaired. Mutations at the conserved tripeptide motif DRY (DRC in case of the OT receptor) result in an either inactive or a constitutively active OT receptor.

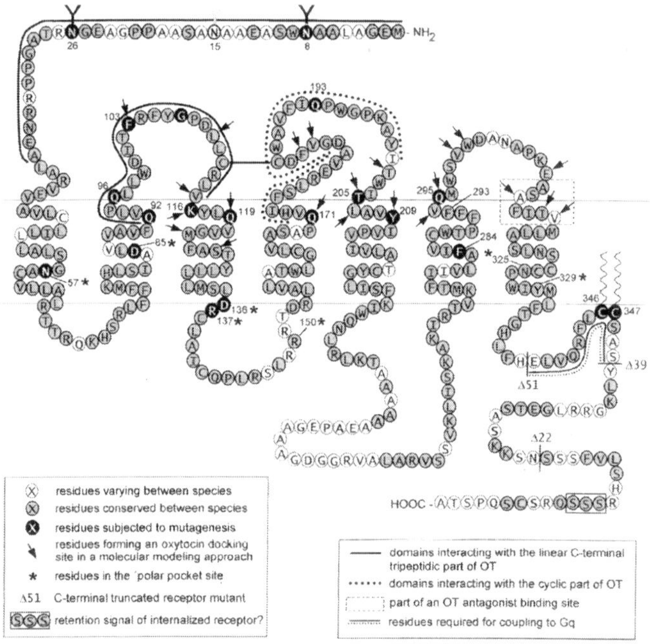

Figure: Schematic model of the human OT receptor indicating amino acid residues that are putatively involved in ligand-binding and associated signal transduction events (described in sect.III, *B–E*). The amino acid residues in gray solid circles are conserved (identical) between the OT receptors from different mammalian species (human, rhesus monkey, pig, bovine, sheep, rat, and mouse). Residues in open circles show interspecies variation. At these positions, an amino acid substitution may be tolerated through mammalian evolution without influencing the functional properties of the receptor. Residues in black solid circles have been subjected to mutagenesis. The glutamine and lysine residues highly conserved within the vasopressin/OT receptor family may partly define an agonist-binding pocket that is common to all the different subtypes of this receptor family. According to a molecular modelling approach, an OT docking site has been proposed (corresponding residues are marked by arrows). In the inactive receptor conformation, the highly conserved arginine may be constrained in a pocket that is formed by polar residues (indicated by asterisks). After agonist binding, this arginine side chain may be shifted out of the "polar pocket," thereby unmasking a G protein binding site. Receptor domains putatively interacting with OT, a peptide OT antagonist and G_qá are marked by lines.

The cysteine residues in the first and second extracellular loops are highly conserved within the GPCR family and are probably connected by a disulfide bridge. Two other well-conserved Cys residues reside within the COOH-terminal domain. Most likely, they are palmitoylated as demonstrated for the V2 receptor and other GPCRs and anchor the cytoplasmic tail in the lipid bilayer. However, for the V2 receptor as well as for the rat OT receptor, elimination of palmitoylation sites by mutagenesis failed to produce significant alterations in receptor function.

The OT receptor has two (mouse, rat) or three (human, pig, sheep, rhesus monkey, bovine) potential *N*-glycosylation sites (N-X-S/T consensus motif) in its extracellular NH_2-terminal domain. For the "core" OT receptor, a molecular mass of 40–45 kDa can be calculated on the basis of the amino acid sequence derived from the known cDNA sequences of several species. In photoaffinity labelling experiments using myometrial membranes obtained from guinea pig during late pregnancy, a 68- to 80-kDa protein was specifically labelled by a photoreactive OT antagonist developed by Elands et al.. Deglycosylation of the photolabelled receptor with endoglycosidase F gave rise to protein with 38–40 kDa. Similarly, in the same tissue, a 78-kDa protein was labelled by a photoreactive vasopressin analogue. In contrast, in membranes from rat mammary gland and rabbit amnion cells, photoreactive OT analogues specifically incorporated into a 65-kDa binding protein. It is possible that the different molecular masses for the myometrial versus the mammary gland and amnion OT receptor are due to differential glycosylation patterns. With the assumption of a mass of 10 kDa for a typical glycosylation core, all of the potential glycosylation sites could be occupied by glycosylation moieties. Recombinant deglycosylation mutants of the human OT receptor have been created by site-directed mutagenesis by exchanging Asp for Asn in positions 8, 15, and 26. The deglycosylated receptors were highly expressed in HeLa cells and showed unaltered receptor binding characteristics. Thus the receptor glycosylation appears not to be necessary for proper expression and has no effect on the functional properties of the receptor. Similar findings have been reported for other GPCRs, e.g., the vasopressin V2 receptor.

Ligand Binding Characteristics C.

The high homology of the nonapeptides of the evolutionary line isotocin-mesotocin-OT is also reflected in the high homology of the corresponding receptors. Accordingly, the mammalian OT receptors share the highest degree of sequence similarity with the toad mesotocin receptor (70%) (6) and the isotocin receptor of teleost fish (66%), whereas the sequence homologies with the vasopressin V1 (nearly 50%) and V2 receptors (40%) are significantly lower. About 100 amino acids (25%) are invariant among the 370-420 amino acids in the human receptors for vasopressin V2, V1a, V1b, and OT. The highest homology between the vasopressin/OT receptor types is found in the extracellular loops and the transmembrane helices. The NH_2 terminus and the COOH terminus have lower similarities, and the intracellular loops are the least of all conserved. Structural common features of the OT/vasopressin receptor family could play an important role in ligand/receptor recognition, e.g., the sequences FQVLPQ at the end of transmembrane domain 2, the sequences GPD (APD in mesotocin receptor) in the first extracellular loop, and DCWA (DCRA and DCWG in mesotocin and isotocin receptor) and PWG in the second extracellular loop.

For small molecules like catecholamines, the ligands bind in a cavity between the α-helical segments formed by transmembrane domains 3–6. Peptide ligands, on the other side, bind more superficially and also interact with extracellular loops and/or the NH_2-terminal domain. For the binding of the peptides OT and arginine vasopressin (AVP), residues located in the transmembrane domains as well as residues within extracellular domains are involved in ligand binding. Because the OT/vasopressin peptides as well as their receptors are well conserved, the ligand binding interaction should consist of both common and selective contact sites. As derived from molecular modelling in combination with mutagenesis studies, the agonist binding site for the vasopressin/OT peptides was proposed to be located in a narrow cleft delimited by the ringlike arrangement of the transmembrane domains. An equivalent position was described for the binding of cationic neurotransmitters. In the rat V1a receptor, the conserved Gln residues in the transmembrane domains 2, 3, 4, and 6 and a Lys residue localised in transmembrane domain 3 were replaced by Ala residues. All the receptor mutants had a decreased affinity for the agonists vasopressin, OT, and vasotocin. Because the corresponding Gln and Lys residues are highly conserved, it was proposed that the agonist-binding pocket is common to all the different subtypes of this receptor family.

Photoaffinity labelling of the first extracellular loop of the bovine vasopressin V2 receptor using a photoreactive lysine vasopressin analogue provided direct evidence for the involvement of the first extracellular loop in agonist binding. In the first extracellular loop, the homologous residues F103, Y115, and D115 in the human OT, V1a/V1b, and V2 receptor were found to be crucial for the determination of the ligand selectivity. For example, the mutation in the equivalent position (Y115F) of the rat V1a receptor led to a 19-fold increase of OT binding compared with the native receptor.

Molecular modelling of ligand binding interaction to the V1a receptor supported the view that the side chain of Arg-8 in AVP projects outside the transmembrane core of the receptor and could interact with Tyr-115 located in the first extracellular loop. Arg-8 in AVP is known to be necessary for its high-affinity binding to the V1a receptor. When Tyr-115 in the V1a receptor is replaced by an Asp and a Phe, the amino acids naturally occurring in the V2 and in the OT receptor subtypes, the agonist selectivity of the V1a receptor switches accordingly. The corresponding residue also determined the agonist specificity of the bovine and pig V2 receptor. Thus this residue certainly contributes to agonist selectivity. Additionally, by a peptide mimetic approach it was found that a synthetic dodecapeptide, which is homologous to the first extracellular loop of the human OT receptor, inhibits the binding of tritiated AVP to the human OT receptor. The second extracellular loop is also thought to be important for hormone binding of the human OT receptor, since it is conserved only within the nonapeptide receptor family.

The OT receptor has a weak ligand selectivity profile: hormones with the same cyclic part and either Arg-8 (in arginine vasotocin) or Leu-8 (in OT) are bound with the same affinity, whereby Ile-3 (in OT) in the cyclic hormone part contributes more to affinity than Phe-3 (in oxypressin). This indicates that the cyclic part of OT is more important in conferring binding selectivity for the OT receptor compared with the linear tripeptidic part of the hormone. Using chimeric "gain in function" V2/OT receptor constructs, Postina et al. demonstrated that the NH_2 terminus and the first and second extracellular loops were necessary for agonist binding and selectivity. In particular, the exchange of the NH_2 terminus of the V2 receptor for the corresponding first extracellular domain of the OT receptor resulted in a sixfold increase in binding affinity for OT. Presumably, the NH_2 terminus of the OT receptor takes part in hormone binding and probably interacts with the hydrophobic leucyl residue in position 8 of the ligands. The NH_2-terminal domain and the first extracellular loop of the OT receptor are proposed to interact with the linear COOH-terminal tripeptidic part of OT, whereas the second extracellular loop of the OT receptor could be identified to interact with the cyclic hormone part. Concerning the binding for OT and AVP, the OT receptor is relatively unselective with only about 10-fold higher affinity of the receptor for OT. AVP acts as a partial agonist on the OT receptor. To elicit the same response as induced by OT, 100-fold higher concentrations of AVP are necessary. However, AVP becomes a full agonist when two aromatic residues of the OT receptor (Y209 and F284) are replaced by the residues F and Y present at equivalent positions in the vasopressin receptor subtypes. These two residues are therefore crucial for the response of the OT receptor to the partial agonist AVP.

Chimeric constructs encoding parts of the white sucker fish [Arg^8]vasotocin receptor and parts of the isotocin receptor have shown that the NH_2 terminus and a region spanning the second extracellular loop and its flanking transmembrane segments contribute to the affinity of the [Arg^8]vasotocin receptor. For the isotocin receptor from teleost fish, it has been shown that the sixth transmembrane helix and/or the fourth extracellular domain are involved in ligand binding.

Several studies indicate that the binding site of OT antagonists is different from the agonist binding site. Studies with chimeric receptors provided evidence that the binding site for the peptide OT antagonist was formed by the transmembrane helices 1, 2, and 7, with a major contribution to binding affinity by the upper part of helix 7. These regions did not participate in OT binding. Most mutations affecting agonist binding affinities have little effect on antagonist binding affinities.

Signal Transduction and G Protein Coupling D

GPCRs may also be constitutively active, in the absence of any agonist. This was first shown for the β-adrenergic receptor where mutations in the third intracellular loop, or simply overexpression of the receptor, resulted in constitutive receptor activation. The human V2 receptor mutant D136A and the human OT receptor mutant R137A represent such constitutively active receptors. Both positions are located within the conserved DRY motif (DRC in the OT receptor) at the cytoplasmic side of transmembrane domain 3. The invariably conserved Arg has been hypothesized to be constrained in a hydrophilic pocket formed by conserved polar residues in transmembrane domains 1, 2, and 7. Receptor activation was suggested to involve protonation of the Asp in this motif causing Arg to shift out of the polar pocket leading to cytoplasmic exposure of buried sequences in the second and third intracellular loops. In accordance with this hypothesis, mutating

Asp in this motif resulted in increased agonist-independent activity of some receptors including the V2 receptor. Although for the V2 receptor as for some other receptors, mutations of the Arg residue within this motif result in uncoupled receptor forms, the human OT receptor mutant R137A possesses an increased basal activity. Thus, in this mutant, conformational constraints are released that normally stabilise the wild-type OT receptor in its inactive ground state. Activation of the OT receptor might occur similarly as proposed for the α_{1B}-adrenergic receptor, i.e., by the opening of a solvent-exposed site in the cytosolic domains that has been hypothesized to be involved in G protein recognition.

OT receptors are functionally coupled to $G_{q/11}\alpha$ class GTP binding proteins that stimulate together with $G\beta^3$ the activity of phospholipase C-β isoforms. This leads to the generation of inositol trisphosphate and 1,2-diacylglycerol. Inositol trisphosphate triggers Ca^{2+}release from intracellular stores, whereas diacylglycerol stimulates protein kinase C, which phosphorylates unidentified target proteins. Finally, in response to an increase of intracellular [Ca^{2+}], a variety of cellular events are initiated. For example, the forming Ca^{2+}-calmodulin complexes trigger activation of neuronal and endothelial isoforms of nitric oxide (NO) synthase. NO in turn stimulates the soluble guanylate cyclase to produce cGMP. In smooth muscle cells, the Ca^{2+}-calmodulin system triggers the activation of myosin light-chain kinase activity which initiates smooth muscle contraction, e.g., in myometrial or mammary myoepithelial cells. In neurosecretory cells, rising Ca^{2+} levels control cellular excitability, modulate their firing patterns, and lead to transmitter release. Further Ca^{2+}-promoted processes include gene transcription and protein synthesis.

In most cell systems studied so far, OT-induced intracellular Ca^{2+} increase is greater in the presence of extracellular Ca^{2+} than that in its absence. This suggests that OT has also effects on calcium influx through voltage-gated or receptor-coupled channels. The effect was nifedipine insensitive. OT was also shown to inhibit Ca^{2+}/Mg^{2+}-ATPase activity in sarcolemmal membranes from the rat uterine myometrium. This could sustain transient increases in intracellular Ca^{2+}concentrations and thereby prolong the effects of OT. In rat, guinea pig, and human myometrial cells, the OT-stimulated phosphoinositide hydrolysis was suggested to be mediated by pertussis toxin-sensitive and/or pertussis toxin-resistant G proteins. OT-stimulated GTPase and phospholipase C activities were attenuated by incubation with an antibody directed against the COOH termini of $G_q\alpha$ and $G_{11}\alpha$ in rat and human myometrial cells. In human myometrium, the coupling of OT receptors to a 80-kDa G protein with transglutaminase activity, termed G_h, was proposed from experiments with solubilised OT receptor-G protein ternary complexes. Phospholipase C-δ1 was suggested as the effector for this kind of signal transduction. In Chinese hamster ovary (CHO) cells expressing the rat OT receptor, OT stimulated increases in intracellular [Ca^{2+}], extracellular signal-related kinase-2 (ERK-2) phosphorylation, and PGE_2 synthesis. OT also induces PGE_2 synthesis in uterine endometrial and amnion cells. In cultured uterine myometrial cells, OT caused tyrosine phosphorylation of mitogen-activated protein (MAP) kinase through an islet-activating protein-sensitive G protein. Solubilisation experiments in combination with pertussis toxin sensitivity assays indicated that rat OT receptors can couple to both $G_{q/11}$ and G_i proteins in transfected CHO cells as well as in pregnant rat myometrium.

Which are the receptor domains conferring G protein specificity? Functional analysis of V1a/V2 hybrid receptors demonstrated that the second intracellular loop of the V1a receptor and the third intracellular loop of the V2 receptor each are required and sufficient for efficient coupling to $G_{q/11}$ and G_s, respectively. Cytoplasmic loops 2 and 3 are also proposed to be implicated in receptor G protein coupling for many other GPCRs. In case of the OT receptor, this appears to be more complex. Several intracellular domains of the receptor could be involved in the specificity and/or efficacy of coupling to $G_{q/11}$. This was concluded from the finding that various coexpressed intracellular receptor domains interfered with the OT-stimulated inositol phosphate production. Hoare et al. provided evidence that proximal parts of the COOH terminus of the rat OT receptor are required for coupling to G_q. Whereas OT receptors with COOH-terminal truncations of 22 and 39 residues showed no effect on receptor function, the OT receptor lacking 51 COOH-terminal residues revealed an interesting phenotype: OT-induced intracellular [Ca^{2+}] transients could be produced, although the phosphoinositide pathway was apparently not activated. However, it remained unclear which signals could mediate this Ca^{2+} release from intracellular stores. The Δ51 mutant receptor had a reduced affinity for OT and was uncoupled from G_q- mediated pathways. A coupling of this receptor to G_i was concluded, since the OT-induced Ca^{2+} transients were sensitive to pertussis toxin and to a $G\beta^3$ sequestrant. Because the Δ39 mutant was still able to couple to both $G_{q/11}$ and G_i, the sequence comprising the residues 339–350 of the rat OT receptor is required for interaction with $G_{q/11}$, but not G_i. Because of the high conservation of the COOH terminus of the OT receptor between various species, similar signal transduction mechanisms may also occur for OT receptors from species other than rats. It is possible that the fidelity of receptor G protein interaction is decreased when OT receptors are strongly upregulated, e.g., in myometrium near term. Moreover, it is known that phosphorylation of GPCRs not only induce their desensitisation but may also modify their coupling specificity.

Receptor Internalisation and Downregulation E

When receptors are persistently stimulated with agonists, they desensitise. This process can occur by numerous mechanisms operating at the transcriptional, translational, and proteïn levels. Rapid, i.e., within seconds to minutes, homologous desensitisation of GPCRs consists of two steps, phosphorylation and subsequent arrestin binding. The receptor uncouples from G proteins and undergoes endocytosis, internalisation, or sequestration. Receptor sequestration is viewed as an early step in the downregulation of receptors that occurs after prolonged (hours to days) agonist stimulation and that may either end in degradation within lysosomes or in recycling back to the plasma membrane. These processes have been best studied for the adrenergic receptors. Birnbaumer and co-workers have analysed the internalisation process for the vasopressin V1a and V2 receptors in some detail.

Like most other GPCRs, OT receptors may undergo rapid homologous desensitisation following persistent agonist stimulation. Within 5–10 min. after agonist stimulation, >60% of the human OT receptors expressed in HEK 293 fibroblasts were internalised (Gimpl and Fahrenholz, unpublished data), similar to as found for the human V2 receptor expressed in the same system and in LLC-PK_1 cells. Internalisation of OT receptors occurs mainly by a clathrin-dependent pathway. But when stably expressed in HEK 293 cells, a fraction of OT receptors (10–15% of total) is localised in caveolae-like membrane microdomains. The internalisation mechanism for this receptor population has not been examined. The internalised OT receptor is not recycled back to the cell surface (Gimpl and Fahrenholz, unpublished data). This indicates that the OT receptor behaves more like the V2 receptor and unlike the V1a receptor that rapidly recycles back to the cell surface. Using mutagenesis experiments and chimeric receptor constructs, Innamorati et al. identified a serine cluster (Ser-362 to Ser-364) in the COOH-terminal tail of the V2 receptor acting as a retention signal for the internalised V2 receptor. The human OT receptor contains 17 potential phosphorylation sites including two serine clusters in its COOH terminus. One may speculate that these clusters also contribute to prevent the recycling of the internalised OT receptor.

Exposure of human myometrial cells to OT for up to 20 h resulted in an almost 10-fold reduction in OT binding capacity. Although the total amount of OT receptor protein appeared not to be affected by OT treatment for up to 48 h, the OT receptor mRNA was reduced, which may be due to transcriptional suppression and/or destabilisation of mRNA. When HEK 293 cells expressing the human OT receptor were treated for 18 h with high (¼M) concentrations of OT, 50% of the initial binding capacity remained at the cell surface. In WRK1 cells, OT was able to induce a desensitisation of the vasopressin (VP) receptors when present for 18 h. Up- or downregulation of receptors could also be affected by yet ill-defined cross-talk mechanisms. Evidence for a cross-talk between the corticotropin-releasing hormone (CRH) and OT signal transduction pathways was provided in human myometrial cells at term. Furthermore, stimulation of β_2-adrenergic receptors causes heterologous upregulation of OT receptors in the nonpregnant estrogen-primed rat myometrium. In this system, a threefold increase in OT receptor mRNA, an 100% rise in receptor binding, and an augmented contractile response of isolated uterine strips to OT were observed.

Effects of Steroids F

Cholesterol 1: Both solubilised and membrane-associated OT receptors require at least two essential components for high-affinity OT binding: divalent cations such as Mn^{2+}or Mg^{2+} and cholesterol. Compared with many other GPCRs, the GTP sensitivity of the agonist binding to the OT receptor is rather modest. All attempts to purify functional OT receptors have been unsuccessful to date. With the use of 3-[(3-cholamidopropyl)dimethylammonio]-2-hydroxy-1-propanesulfonate (CHAPSO) as detergent, it is possible to solubilise functional OT receptors from different sources. However, a common observation is that following solubilisation, OT receptors lose characteristic binding properties, the affinity for OT becomes lower, and/or additional low-affinity state receptors appear in the extract. Unfortunately, low-affinity [e.g., dissociation constant (K_d) >10–50 nM] receptor populations cannot be characterised adequately by conventional radioligand binding assays. The necessity to separate free from bound ligand concomitantly leads to a dissociation of low-affinity ligand-receptor interactions. Solubilisation with CHAPSO is known to lead to a substantial cholesterol depletion of the soluble extract, and we could demonstrate that substitution with cholesterol markedly enhanced the OT binding of soluble OT receptors. This became first evident in reconstitution of soluble OT receptors using liposomes of defined composition. A saturable high-affinity OT binding was obtained only with liposomes that contained a critical amount of cholesterol. Moreover, when OT receptors were expressed in insect cells, which naturally have plasma membranes with low cholesterol content, the receptors are mainly in a low-affinity state (K_d >100 nM). After addition of cholesterol to the culture medium, a fraction of OT receptors is converted from a low- to a high-affinity state (K_d 1 nM). The low-affinity state was identified as a physiological active receptor state, and the conversion of the affinity states to each other is, at least to a certain degree, reversible. The interaction of cholesterol with OT receptors is of high specificity and is not due to mere changes of membrane fluidity. Furthermore, cholesterol stabilises both membrane-associated and solubilised OT receptors against thermal denaturation. Taken together, our data suggest a direct and cooperative molecular interaction of cholesterol with OT receptors. Cholesterol acts as an allosteric modulator and stabilises the receptor in a high-affinity state for agonists and antagonists. In many but not in all cell systems, populations of high- and low-affinity OT receptors have been observed. This could reflect uneven cholesterol distributions within the plasma membrane of these cells. We would expect that high-affinity state OT receptors are preferentially localised in cholesterol-rich subdomains of the plasma membrane. Likewise, receptor heterogeneity with respect to affinity states should be highest in cell systems with abundant cholesterol-rich domains such as rafts or caveolae structures, e.g., in myometrial cells at term. In fact, we recently provided evidence for a partial enrichment of high-affinity OT receptors in cholesterol-rich plasma membrane domains in HEK 293 fibroblasts stably expressing the human OT receptor.

Divalent metal ions like Mg^{2+} are long known to increase the response of target cells to OT and to shift the dose-response curve to the left. Thus addition of Mg^{2+} was found to increase both the OT binding capacity and the affinity state of the OT receptor. This is surprisingly similar to what cholesterol does. In addition, Mg^{2+} has been proposed to display its effect on the OT receptor interaction by influencing positive cooperativity. Mg^{2+} increases the potency of OT analogues in stimulating uterine contractions whereby the effects of Mg^{2+} were observed to be inversely related to the potency of the peptide.

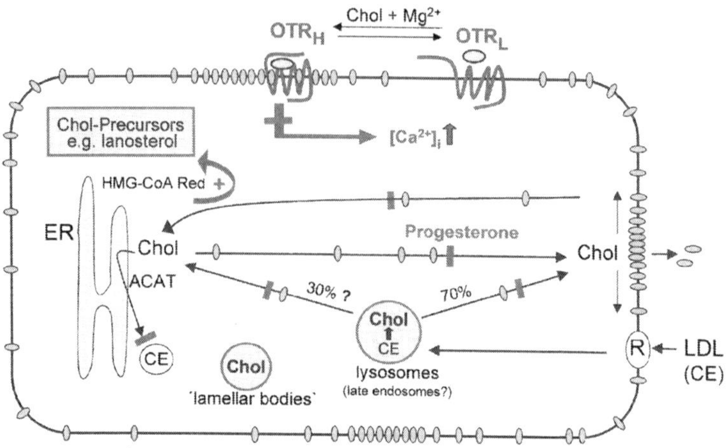

Fig. 6. Schematic model of nongenomic inhibitory effects of progesterone. Progesterone inhibits both the signal transduction of Gq couplled receptors (as shown here for the OT receptor) and the intracellular trafficking of cholesterol. Principally, eukaryotic cells can obtain the required cholesterol (Chol, gray ellipses) by two sources: endogenously by de novo synthesis of cholesterol and exogenously by uptake of cholesteryl ester (CE)-rich low-density lipoprotein (LDL) particles via receptor-mediated (R) endocytosis. De novo synthesized cholesterol first arrives at cholesterol-rich domains in the plasma membrane (caveolae and/or "lipid rafts") that may function as cholesterol "sorting centres" within the plasma membrane, where most of the cellular cholesterol resides. Progesterone blocks several intracellular transport pathways of cholesterol (red bars) except for the LDL receptor-mediated uptake of cholesterol. Moreover, cholesterol esterification does not occur in the presence of progesterone, presumably due to the lack of cholesterol substrate for acyl-CoA:cholesterol acetyltransferase (ACAT). As a consequence, unesterified cholesterol accumulates in lysosomes (or late endosomes) and lysosome-like compartments (designated as "lamellar bodies") (marked by red background). The key enzyme for the cholesterol de novo synthesis, 3-hydroxy-3-methylglutaryl-CoA reductase (HMG-CoA Red), is stimulated in the presence of progesterone (red arrow), but the cholesterol biosynthesis stops at the level of precursors (e.g., lanosterol). Enzymes involved in the conversion of cholesterol precursors reside in the endoplasmic reticulum (ER), and progesterone most likely prevents sterol precursors localised in the plasma membrane from reaching the ER-resident enzymes, thereby preventing their conversion to cholesterol. Overall, progesterone induces a state of cholesterol auxotrophy. However, after progesterone withdrawal, the accumulated precursors will be rapidly converted to cholesterol. Thus cells will become overloaded with cholesterol for a certain period of time, after which the cholesterol homeostasis will be reestablished. We hypothesize that these reversible progesterone-induced changes of the cholesterol trafficking could have a strong influence on signal transduction processes, particularly in case of the OT receptor (OTR_H and OTR_L, high-affinity and low-affinity OT receptor, respectively; receptor in blue; OT in yellow).

Relatively inactive peptides like 7-glycine OT became significantly more potent when the Mg^{2+} concentration bathing the uterine smooth muscle in vivo was increased from 0 to 0.5 mM. Conclusively, cholesterol and Mg^{2+} are essential allosteric modulators of the OT receptor and may be involved in the regulation of OT-mediated signalling functions.

Do these allosteric modulators play a role for the regulation of OT-related physiological processes? Some reports suggest that particularly in reproductive tissues, the cholesterol concentrations may be highly dynamic. With the use of freeze-fracture cytochemistry with the cholesterol-binding filipin, marked increases in cholesterol have been found in rat uterine epithelial cells at the time of blastocyst implantation. In the human placental syncytiotrophoblast basal membrane, Sen et al. observed a steady decrease in cholesterol-to-phospholipid ratio in correlation with an increase in membrane fluidity during placental development. At term however, the cholesterol-to-phospholipids ratio in syncytiotrophoblast membranes was found to be increased compared with the cholesterol-to-phospholipid ratio in early placentas. Moreover, cholesterol-enriched caveolae structures are a conspicuous feature in the rat myometrium at term. We have provided evidence that cholesterol can modulate receptor function by both changes of the membrane fluidity and direct binding effects, e.g., in case of the OT receptor. Plasma membranes with lowered cholesterol content showed a decreased capacity (B_{max}) of binding sites and/or a decreased affinity (K_d) of ligand-receptor binding. Interestingly, Lopez et al. reported that pregnancy in humans was associated with increases in both density and affinity of OT receptors. To draw further conclusions, correlation studies are required using tissues in which both the membrane cholesterol content and the OT receptor activity will be measured at the same time.

Progesterone 2

Progesterone is considered to be essential to maintain the uterine quiescence. Grazzini et al. recently postulated that progesterone specifically binds to the rat OT receptor with high affinity (K_d 20 nM) and thereby inhibits the receptor function. In case of the human OT receptor, a direct inhibitory interaction [inhibitory constant (K_i) 30 nM] with a progesterone metabolite, 5β-pregnane-3,20-dione, has been reported by the same authors. They claimed that progesterone could act as a negative modulator of the OT receptor and thus offered a plausible mechanism of how progesterone could contribute to uterine quiescence. However, these findings could not be reproduced in several other laboratories including our own. Instead, we found that high concentrations of progesterone (>10 ¼M) attenuated or blocked the signalling of several GPCRs, including the OT receptor. The progesterone effects occurred within minutes, were reversible, and could not be blocked by a protein synthesis inhibitor. Overall, the action of progesterone was more cell type specific than receptor specific. The progesterone doses that are required to affect the signalling function of receptors are much higher than the progesterone levels found in plasma or in nonsteroidogenic tissues such as the myometrium. In steroidogenic tissues, however, huge amounts of progesterone have been measured.

In human corpus luteum, progesterone concentrations reached peak levels of 25 ìg/g tissue shortly after ovulation and in the early luteal phase. These values are within the range of the progesterone concentrations that were effective in our study. Thus, in steroidogenic cells as well as in their environment, progesterone might nongenomically influence the signalling of receptors. The molecular mechanisms underlying this progesterone action are not understood. A well-known progesterone binding protein is the multidrug resistance P-glycoprotein.

In addition to their role in detoxification, P-glycoproteins are involved in intracellular cholesterol transport. It is known that progesterone markedly interferes with the intracellular transport (and metabolism?) of cholesterol. At concentrations in the micromolar range, it inhibits both the cholesterol esterification and the transport of cholesterol to and from the plasma membrane. In particular, progesterone reduces the cholesterol pool residing in caveolae . Paradoxically, at the same time, progesterone stimulates the activity of 3-hydroxy-3-methylglutaryl (HMG)-CoA reductase, the key enzyme of de novo cholesterol biosynthesis. Hence, cholesterol precursors like lanosterol begin to enrich in the membranes of the cell. As mentioned above, the OT receptor needs a cholesterol-rich microenvironment to become stabilised in its high-affinity state. Because the cholesterol precursors, particularly lanosterol, are completely inactive to support the OT receptor in its high-affinity state, the responsiveness of the OT system may not be fully operative during the continuous presence of high progesterone concentrations. According to this scenario, progesterone withdrawal would restore the cholesterol transport so that the highly enriched amounts of cholesterol precursors would now become rapidly converted to cholesterol. This would lead to a sudden rise of cholesterol and should push the responsiveness to OT since low-affinity OT receptors could now be converted into their high-affinity state. According to this postulated mechanism, progesterone could affect the signalling of all those receptors that are functionally dependent on cholesterol. It is important to note that the nongenomic actions of progesterone including its influence on the cholesterol transport require progesterone concentrations in the micromolar range. This suggests that the described effects may be limited to the steroidogenic tissues and to their environment. Most likely, progesterone acts in these tissues via both genomic and nongenomic pathways together with other steroids to control receptor activity.

The Peripheral Oxytocin System

Female Reproductive System A

Uterus 1: The pregnant uterus is one of the traditional targets of OT. OT is one of the most potent uterotonic agents and is clinically used to induce labour. Accordingly, the development of highly specific OT antagonists may be of therapeutic value for the prevention of preterm labour and the regulation of dysmenorrhea.

The OT gene was found to be expressed in the rat uterine epithelium at term. The estrogen-induced elevation of OT mRNA levels was restricted to 3 days and reached peak levels that exceeded hypothalamic OT mRNA levels by a factor of 70. In rats, OT gene expression was shown to be present in placenta and amnion and in humans in amnion, chorion, and decidua. However, in most studies, significant increases of OT before the onset of labour have not been detected, neither in maternal plasma nor in intrauterine tissues. On the other hand, some findings suggest that there is a relationship between the pattern of OT secretion and advancing pregnancy. In rhesus monkeys, it was shown that maternal but not fetal OT concentrations were positively correlated with nocturnal uterine activity and progressively increased during late pregnancy and delivery.

Around the onset of labour, uterine sensitivity to OT markedly increases. This is associated with both an upregulation of OT receptor mRNA levels and a strong increase in the density of myometrial OT receptors, reaching a peak during early labour. This has been demonstrated both in the rat and in the human species, in which receptor levels rise during early labour to 200 times that in the nonpregnant state. Thus, at the onset of labour, OT can stimulate uterine contractions at levels that are ineffective in the nonpregnant state. After parturition, the concentrations of OT receptors rapidly decline. In rats, the uterine OT receptor mRNA levels decreased more than sevenfold within 24 h. Possibly, the downregulation of the OT receptors may be necessary to avoid unwanted contractile responses during lactation when OT levels are raised.

Gonadal steroids play an important role in the regulation of uterine OT receptors. In the days preceding birth, the ratio of plasma progesterone to estrogen falls. These changes in the steroid concentrations may occur in most mammals. At least in humans, progesterone withdrawal has not been determined. The steep drop of circulating progesterone occurs after placental delivery. Alteration of sex steroid metabolism in the fetoplacental unit appears to occur in women and primates. As shown in bovine and in sheep, maturation of the fetal hypothalamus leads to an increased secretion of CRH, which in turn stimulates the pituitary to secrete ACTH. Subsequently, ACTH stimulates the fetal adrenal to release cortisol.

Additionally, OT-induced contractures of the myometrium in late pregnancy could lead to temporary decreases in blood flow and transient episodes of fetal hypoxia, which also may provoke a fetal stress response. Cortisol then increases the activity of the key enzyme cyctochrome *P*-450c17α, which promotes placental pregnenolone turnover into estrogens. In primates, CRH is additionally synthesized by the placenta and stimulates the fetal adrenal to secrete dehydroepiandrosterone sulfate (DHEAS), the precursor of placental estrogen.

In each case, the ratio of estrogen to progesterone increases in the maternal plasma concomitantly with increases in the synthesis of connexin-43, formation of gap junctions, increased production of prostaglandins from intrauterine tissues, and an upregulation of OT receptors. Finally, the uterine quiescence that was maintained by the high progesterone level ceases, and parturition can occur. Upregulation of OT receptors and an increased expression of gap junctions also occur in the hours or days preceding human labour onset. Obviously, estrogens and progesterone act in opposing fashion on the function, expression, and/or regulation of OT receptors. The mechanisms for the sudden and sometimes unpredictable responsiveness to OT of the myometrium is still mysterious. As Kimura pointed out, the reaction of the uterus to OT in the same patient could vary from day to day. To induce labour at term, in some patients, OT is completely ineffective even at high doses, whereas in others a minimal dose can induce hypertonus of the uterus. However, despite the striking steroid dependence of the OT system, the promoter of the human OT receptor gene does not contain a classical steroid responsive element. Several observations indicated that progesterone promotes uterine relaxation and inhibits the function of the OT receptor system by both genomic and nongenomic mechanisms. Even in the absence of protein synthesis, progesterone induces a reduction in uterine OT binding. Moreover, progesterone-mediated downregulation of OT receptors was not accompanied by a decrease in OT receptor gene expression. The mechanisms of how progesterone acts nongenomically are still unknown but are of fundamental importance.

The increase in OT receptors before labour is not confined to the myometrium. It is also found in the decidua. During the course of parturition, the OT receptor gene expression in human chorio-decidual tissue was fivefold increased. In the decidua, OT has a separate action of stimulating the release of prostaglandin $PGF_{2\alpha}$. At the end of pregnancy, an increased secretion of $PGF_{2\alpha}$ drives luteolysis and thus leads to progesterone withdrawal and labour initiation in rodents. Mice lacking the gene encoding the $PGF_{2\alpha}$ receptor developed normally but were unable to deliver normal fetuses at term. These knock-out mice neither responded to OT nor did they show an induction of OT receptors. Furthermore, the normal decline of serum progesterone concentrations that precedes parturition did not occur. Ovariectomy restored induction of OT receptors and permitted successful delivery in the $PGF_{2\alpha}$ receptor-deficient mice. This indicates that $PGF_{2\alpha}$ acts upstream of OT to induce luteolysis, i.e., production of progesterone. The OT system may be regarded as a key regulator of parturition, with the induction of OT receptors as a trigger event. However, in OT-deficient mice, parturition remained unaffected. This suggests that at least in mice, uterotonins other than OT are (additionally?) operative as myometrial contractants to initiate labour. A recent study with mice deficient in both OT and cyclooxygenase-1 shed more light for OT's role at the onset of labour. Cyclooxygenase-1 is involved in the synthesis of prostaglandins, and as expected, mice lacking this enzyme showed reduced levels of $PGF_{2\alpha}$, impaired luteolysis, a less pronounced fall in progesterone in late gestation, and finally a delayed initiation of labour. Surprisingly, mice deficient in both OT and cyclooxygenase-1 initiated labour at the normal time. Thus OT and prostaglandins had apparently opposing actions on the onset of murine parturition: a luteotrophic function of OT versus a luteolytic action of prostaglandins. So, OT stabilises the progesterone synthesis before parturition. For the normal timing of labour it might be essential to generate sufficient $PGF_{2\alpha}$ to overcome the luteotrophic action of OT in late gestation. These findings might explain why the upregulation of OT receptor occurs without an increased expression of OT shortly before the onset of labour. The induction of uterine OT receptor expression allows OT to act as a potent uterotonic agent, whereas an increase in OT expression could prolong gestation due its trophic effects on the corpus luteum. Fuchs et al. reported that OT may also be involved in induction of cyclooxygenase-2 expression and $PGF_{2\alpha}$ release at term and during parturition in cows.

In ruminants, the endometrial OT receptor exhibits a cycle-dependent regulation and plays a crucial role in reproduction. In the bovine endometrium, OT receptor mRNA levels can exceed the levels in the myometrium at the same stage of the cycle. Within 2–3 days around estrus, the endometrial OT receptors increase to a level similar to that observed at term of pregnancy. For the remainder of the cycle, OT receptor concentrations are almost undetectable. Pulsatile secretion of endometrial $PGF_{2\alpha}$ is stimulated by OT during late diestrus in domestic ruminants and results in regression of the corpus luteum leading to the onset of a new estrous cycle. Thus the uterus influences the length of the luteal phase via the expression of the endometrial OT receptor. The administration of an OT antagonist or the continuous administration of OT, which downregulates the OT receptor, delays the onset of luteolysis and lengthens the cycle. A steep rise in endometrial OT receptors also precedes the onset of labour in ruminants. In cows, the receptor density increases almost 200-fold and remains at high levels during labour. Suppression of the endometrial OT receptor gene expression was shown to be caused by antiluteolytic proteins, e.g., interferon-[3] secreted by the trophoblast of the developing blastocyste. In primates, the endometrium is not essential for luteolysis, since hysterectomy does not abolish cyclic ovarian function. In humans, Northern blotting revealed the presence of OT receptor mRNA in the pregnant myometrium, in the endometrium, and in the ovary. In the nonpregnant human uterus, the OT receptor mRNA is mainly expressed in the glandular epithelial cells of the endometrium with highest expression level occurring at ovulation. However, its physiological role is unclear.

One of the highest concentrations of OT receptors (10 pmol/mg protein) has been measured in rabbit amnion at term. At the end of gestation, rabbit amnion OT receptors increase more than 200-fold. This receptor upregulation is associated with the release of PGE_2 from cultured amnion cells and suggests an important role for the amnion and OT in the initiation of labour in rabbits. In the same system, cortisol and cAMP caused a marked physiological upregulation of OT receptors as well as an increased PGE_2 response to OT. Addition of cortisol to amnion cells increased OT-stimulated PGE_2 release almost 100-fold, whereas the combination of forskolin and cortisol increased the PGE_2 response to OT 5,600 times. OT receptors in human amnion were also associated with PGE_2 release and were found to be upregulated in early and advanced labour, albeit to a much lower degree. In contrast, no OT binding sites were observed in amniotic/chorionic or placental membranes from rats.

Many studies have supported the view that OT is synthesized in a paracrine system operating within human intrauterine tissues. This includes the fetal membranes amnion and chorion and the maternal decidua. These tissue layers are continuously bathed in amniotic fluid and could transmit signals of maternal or fetal origin to the myometrium. Because fetal membranes are capable of steroidogenesis, they may contribute to local changes in the estrogen-to-progesterone ratio. Although there is evidence that during late gestation OT gene expression is upregulated in intrauterine tissues, the OT peptide itself was not found at concentrations significantly higher than in the circulating blood. This could be due to the high activity of oxytocinases, but that still remains to be shown. The possibility of local, paracrine effects of OT was also suggested to be present within the intrauterine tissues of the rat and bovine. In the pregnant cow, OT mRNA levels were found to be very low except for the corpus luteum. After the onset of labour, both OT gene and OT peptide were expressed at significant levels in the corpus luteum. This rescue of luteal OT at term could act to supplement the circulating hormone of pituitary origin. Thus OT may act primarily as a local mediator and not as a circulating hormone during parturition. In an autocrine/paracrine system within the uterus, significant changes of OT, prostaglandins, and sex steroids could occur without being reflected in the maternal circulation.

Ovary and Corpus Luteum 2

In several species, the ovary has been shown to contain OT and may be a site of local OT production. In the marmoset monkey *(Callithrix jacchus)*, in vivo and in vitro studies suggested that OT is a follicular luteinization factor. After human chorionic gonadotropin treatment, almost all granulosa cell layers in antral follicles showed immunoreactivity for both OT and OT receptor. OT was produced only by granulosa cells derived from preovulatory follicles, and after application of OT, only the granulosa cells cultured from preovulatory follicles elicited an increase in progesterone production. Functional OT receptors have been detected in bovine granulosa cells, suggesting that OT may be an autocrine factor during follicular growth. OT also increased the rate of mouse blastocyst development and might therefore play some role in the early stage of development of fertilized oocytes. Both OT and OT receptor genes are expressed in human cumulus cells surrounding the oocytes. Thus local OT may participate in fertilization and early embryonic development in humans.

The interaction of neurohypophysial OT with endometrial OT receptors evokes the secretion of luteolytic pulses of uterine $PGF_{2\alpha}$. McCracken et al. have postulated the concept of a central OT pulse generator that functions as pacemaker for luteolysis.

According to this concept, the uterus transduces hypothalamic signals in the form of episodic OT secretion, into luteolytic pulses of uterine $PGF_{2\alpha}$. In ruminants, luteal OT is released synchronously by uterine $PGF_{2\alpha}$ pulses. Thus a positive-feedback loop is established that amplifies neural OT signals.

The onset of this feedback loop and episodic $PGF_{2\alpha}$ secretion is controlled by the appearance of OT receptors in the endometrium. During pregnancy, the developing conceptuses must prevent endometrial $PGF_{2\alpha}$ from being released into the uterine vasculature to prevent corpus luteum regression and progesterone withdrawal. The continued progesterone secretion is required for establishment and maintenance of pregnancy.

Luteal cells produce OT, but they can also be target cells for OT action. OT receptors have been characterised on both steroidogenic cell types of the corpus luteum, small and large cells. In several species, OT release was highest in the young corpus luteum. In the pig, the density of OT receptors was also highest during the early luteal phase, suggesting that autocrine/paracrine actions of OT may occur primarily in the young corpus luteum. The effects of OT on cultivated luteal cells were also strongly dependent on the age of the corpus luteum. In the pig, a function for OT in the ovary has been suggested in both a luteotrophic and luteolytic context. Luteal OT and progesterone release occurs in tightly coupled pulses. In vivo, OT and $PGF_{2\alpha}$ stimulate estradiol and progesterone release, and estradiol itself further stimulates progesterone release. Thus there may be an intraluteal circuit that involves paracrine effects of estradiol, OT, and $PGF_{2\alpha.}$

Male Reproductive Tract B

Testis 1

In several species, a pulse of systemic OT, presumably of hypothalamic origin, appears to be associated with ejaculation. The systemic hormone could act peripherally stimulating smooth muscle cells of the male reproductive tract, but could also reflect central effects in the brain modulating sexual behaviour.

OT has been identified in the testes from various mammalian species, and a mesotocin-like peptide has been demonstrated in the testes of birds and marsupials. There is evidence that OT is made locally within the testis, and possibly also the epididymis and prostate. Concerning the localisation of the OT system in the male reproductive tract, species-specific differences are important to note. For example, mice do not have detectable levels of OT mRNA in their testes, whereas cattle have relatively high levels of testicular OT mRNA. In contrast to the eutherian mammals, in the tammar wallaby (*Macropus eugenii*), the mesotocin receptor gene and protein, which are highly homologous to the eutherian OT receptor, are expressed in the prostate gland, but not in the testis. In the human, the complete OT system appears to be present in testis, epididymis, and prostate.

In the rat testis, OT concentrations are higher than those in the peripheral circulation. OT gene transcripts could be detected using PCR analysis but not by Northern hybridisation. Within the rat testes, OT is present in the interstitial Leydig cells, the cells which provide the main source of testosterone in the male. Testicular OT levels are not constant but vary in dependence of the level of gonadotrophins and the activity of the seminiferous epithelium. OT production is increased in vitro by luteinizing hormone (LH), whereas testosterone itself has no effect on the secretion of OT. Notably, OT was only detectable during active spermatogenesis. Primarily two functions have been ascribed to testicular OT, namely, the regulation of seminiferous tubule contractility and the modulation of steroidogenesis. In the testis, the seminiferous tubules are surrounded by smooth muscle-like cells, the myoid cells. OT has been shown to enhance the contractility of the tubules; therefore, the responsiveness to OT was higher at a certain stage of the spermatogenic cycle, around the time when the sperm are shed into the lumen. Obviously, OT promotes the spermiation and the subsequent transport of the immotile spermatozoa to the epididymis. However, there is no clear evidence for the expression of OT receptors on myoid cells, the cells which are mainly responsible for the contractile activity of the tubules. Because Sertoli cells exhibit the components of a local OT system under some circumstances, they may also be involved in the contractile activity of seminiferous tubules. On the other hand, it was suggested that the contractility effects are mediated by V1a receptors that are present in testes at severalfold higher densities compared with OT receptors.

Uterine-type OT receptors have been identified in the interstitial spaces in the rat testis consistent with binding to Leydig cells. Also in the male marmoset monkey (*Callithrix jacchus*), OT and its receptor have been detected predominantly within the Leydig cells. The Leydig cells drive spermatogenesis via the secretion of testosterone, which acts on the Sertoli and/or peritubular cells to create an environment that enables normal progression of germ cells through the spermatogenic cycle. Daily subcutaneous injections of OT led to an increase in the plasma and testicular levels of testosterone. In contrast, continuous administration of the peptide into the testis, whether by implants or in the transgenic mouse which overexpresses the bovine OT gene in its testes, produced a decrease in testosterone but an increase in dihydrotestosterone concentrations. Although in the testis the predominant androgen is testosterone, elsewhere in the male reproductive tract testosterone acts as a prohormone and is converted to its active metabolite dihydrotestosterone by the enzyme 5α-reductase. OT was shown to increase the activity of 5α-reductase in both testis and epididymis and may thus have an autocrine/paracrine role modulating steroid metabolism in these tissues.

Prostate Gland 2

The prostate is an androgen-dependent gland. Testosterone enters the cell and is converted to dihydrotestosterone, which then modulates growth and prostatic functions. OT is present in the prostate at concentrations higher than in the plasma and can increase the resting tone of prostatic tissue from guinea pig, rat, dog, and human. OT also evoked contractile activity on mammalian prostates in vitro. Thus it was suggested that OT is involved in the contraction of the prostate and the resulting expulsion of prostatic secretions at ejaculation.

OT may also be involved in the pathophysiology of benign prostatic hyperplasia. In humans, benign prostatic hyperplasia occurs spontaneously, is often associated with bladder outlet obstruction, and affects 50% of men over the age of 60. However, obstructive symptoms may not only be caused by an enlargement of the gland but also by an increase in smooth muscle tone. So, in the nonoperative management of prostatic hyperplasia, α_1-adrenergic blockers are presently used. Because OT is even more potent at increasing smooth muscle tone than norepinephrine, OT antagonists may be potentially useful in the treatment of this disease.

OT can stimulate growth of the prostate in the rat, particularly the mitotic activity in the glandular epithelium. OT treatment elevated the testosterone levels of plasma, testes, and prostate. Conversely, prostatic OT concentrations are found to be decreased by testosterone and increased after castration. In rats, OT treatment only transiently increased 5α-reductase activity in the prostate, consistent with the short-lived increase of dihydrotesterone in this tissue. These observations suggest that local feedback mechanisms act to control prostatic levels of dihydrotestosterone and prostatic growth. The findings in rats were somewhat different to those in dogs, which unlike rats can spontaneously develop prostatic hyperplasia. Interestingly, old dogs with established prostatic hyperplasia showed raised OT levels accompanied by elevated 5α-reductase activity within their prostatic tissues. It is hypothesized that OT acts as a paracrine factor to regulate cell growth via its influence on the enzyme 5α-reductase.

Mammary Tissues C

Milk Ejection 1: One of the classical roles assigned to OT is milk ejection from the mammary gland. The secretion of the mammary glands is triggered when the infant begins to suck on the nipple. The stimulation of tactile receptors at that site generates sensory impulses that are transmitted from the nipples to the spinal cord and then to the secretory oxytocinergic neurons in the hypothalamus. These neurons display a synchronised high-frequency bursting activity, consisting of a brief (3–4 s) high-frequency discharge of action potentials recurring every 5–15 min. Each burst leads to massive release of OT into the bloodstream by which OT is carried to the lactating breasts. There it causes contraction of the myoepithelial cells in the walls of the lactiferous ducts, sinuses, and breast tissue alveoli. In humans, within 30 s to 1 min. after a baby begins to suckle the breast, milk begins to flow. This process is called milk ejection or milk let-down reflex and continues to function until weaning.

As mentioned above, central oxytocinergic neurons are decisive components for the initiation and maintenance of successful lactation. So, the reflex that elicits OT release in women is activated before the tactile stimulus of suckling occurs and is related to such factors as the baby crying. The essential role of OT for the milk let-down reflex has been confirmed in OT-deficient mice. In fact, the major deficit of these female OT knock-out mice was the failure to nurse their offspring. In addition, the presence of OT in conjunction with continued milk removal was also required for postpartum alveolar proliferation and mammary gland function. Alveolar density and mammary epithelial cell differentiation at parturition were similar in wild-type and OT-deficient dams. However, within 12 h after parturition, 2% of the alveolar cells in wild-type dams incorporated DNA and proliferated, whereas no proliferation was detected in OT-deficient dams. Continuous suckling of pups led to the expansion of lobulo-alveolar units in wild-type but not in OT-deficient dams. Despite suckling and the presence of systemic lactogenic hormones, mammary tissue in OT-deficient dams partially involuted. Continuously elevated OT concentrations such as those during infusion or during normal milking are also necessary for complete milk removal in diary cows.

Biochemical studies showed that the mammary OT receptor is a 65-kDa protein (in rats), coupled to phosphoinositide turnover, and is most likely the same as the uterine-type OT receptor. In rats, Soloff and Wieder reported an ~100-fold increase in the number of OT receptors per mammary gland between the first day of pregnancy and late lactation. A very strong increase in OT binding sites toward term was also observed in porcine mammary tissues. In humans, however, Kimura et al. did not find an elevated expression level of the OT receptor mRNA during lactation. As judged by OT receptor immunoreactivity, unexpectedly, the ductal/glandular epithelium revealed a higher density of OT receptors compared with myoepithelial cells in both the lactating and nonlactating breast in humans and in the common marmoset (*Callithrix jacchus*).

Breast Cancer and Tumor Cells 2

Breast cancer is the leading cause of death in women between ages 35 and 45, but it is most common in women over age 50. It is unknown why mothers who breast-fed their babies have a 20% lower incidence of breast cancer after menopause than mothers who did not. Murrell proposed a hypothesis that partially links breast cancer to the activation of the OT system. Accordingly, carcinogens in the breast may be generated by the action of superoxide free radicals released when acinal gland distension causes microvessel ischemia. Thus inadequate nipple care in the at-risk years could lead to ductal obstruction preventing the elimination of carcinogens from the breast.

The regular production of OT from nipple stimulation would cause contraction of the myoepithelial cells, relieving acinal gland distension and aiding the active elimination of carcinogenic fluid from the breast. Thus OT production was suggested as a preventative factor in the development of breast cancer both pre- and postmenopausally.

OT receptors have been described in a number of human breast tumours and breast cell lines, e.g., the human tumor cell lines MCF-7 or Hs578T. Copland et al. recently studied the behaviour of OT receptors in dependence of culture conditions in Hs578 cells. Hs578 cells responded to OT by an increase of intracellular $Ca^{2+}$and stimulation of ERK-2 phosphorylation and PGE_2synthesis. OT receptors in these cells were strongly downregulated by serum starvation. Conversely, restoration of serum and addition of dexamethasone (1 ¼M) increased the OT receptor levels by nearly 10-fold and also led to an elevation of the steady-state OT receptor mRNA level. These cells may be good models for future studies with respect to OT receptor regulation in dependence of steroids.

Controversial results have been published concerning a proliferative action of OT on breast tumor cells. Ito et al. found that OT had no effect on the growth of different breast cell lines cultured for 7 days. The OT receptor was expressed in breast cancer derived not from the myoepithelium but from the glandular or ductal epithelium. According to Sapino et al., OT exerts a trophic effect on myoepithelial cells in the mammary gland. In organotypic cultures of the mouse mammary gland, OT induced differentiation and proliferation of myoepithelial and, to a lesser extent, luminal epithelial cells. On the other hand, OT inhibited cell proliferation and tumor growth of rat and mouse mammary carcinomas. In MCF7 cells for instance, OT inhibited estrogen-induced cell growth and enhanced the inhibitory effect of tamoxifen on cell proliferation. Antiproliferative effects of OT have also been observed in various breast carcinoma cell lines as well as in human neuroblastoma and astrocytoma cells, and these effects were accompanied by the activation of the cAMP/protein kinase A pathway. In vivo, OT reduced the growth of certain mammary tumours. With the use of immunohistochemistry and RT-PCR, OT receptors and its corresponding mRNA were detected in normal and pathological breast tissues. Interestingly, the expression of OT receptors was positively correlated with the progesterone level of these tissues. Further studies are required to clarify the role of the OT system concerning the long-term effects in mammary tissues and tumor cell lines derived from them.

Kidney D

The kidney is one of the peripheral target tissues for neurohypophysial hormones that exercise control over hydromineral excretion. The hormones are released into the blood by stimulations, such as hypovolemia or hyperosmolarity, that generally activate the OT and AVP neurons. When plasma sodium concentration exceeds 130 mM, the levels of both hormones increase as an exponential function of plasma sodium concentration. OT is a nonhypertensive natriuretic agent. It is involved in normal osmolar regulation, which is presumably different from the volume regulatory components of Na^+ homeostasis. Acute administration of OT to conscious rats produced a modest increase in the glomerular filtration rate and effective filtration fraction. The natriuretic effect of OT is mainly due to a reduction in tubular Na^+ reabsorption, probably in the terminal distal tubule or the collecting duct.

Autoradiographical analysis showed the existence and precise localisation of OT receptors in the rat kidney. Interestingly, the distribution of OT binding sites undergoes reshaping during postnatal development as it was similarly observed in the rat brain during maturation. Specific OT binding sites have been first detected at embryonic *day 17* in the cortex. In the medulla, OT binding sites were first detected at embryonic*day 19* when this region is forming. In the adult rat, OT binding sites were exclusively localised in the cortex, mainly concentrated in the macula densa. In the inner medulla, the OT binding sites were found on the loops of Henle of the juxtamedullary nephrons. Interestingly, at all stages examined, cortical OT-binding sites had a higher selectivity for OT versus vasopressin compared with the medullary sites. It was therefore suggested that OT binding sites of the macula densa and thin Henle's loop could represent two subtypes of OT receptors. The location profile of the receptors suggests a possible role for OT in the regulation of tubuloglomerular feedback and solute transport.

Estrogen induced OT receptor gene expression in the outer medulla region and increased expression of OT receptors in macula densa cells of ovariectomised female and adrenalectomised male rats. Experiments with the antiestrogen tamoxifen suggested cell-specific regulation of OT receptor expression in macula densa and proximal tubule cells. Thus OT receptors may mediate estrogen-induced alterations in renal fluid dynamics. OT receptor mRNA levels have been measured in kidneys of late-pregnant, peri-parturient, and lactating rats. Of interest,

OT receptor transcripts could not be detected in renal tissues of peri-parturient females, at times when OT receptor mRNA levels were highest in uterus. However, OT receptor gene expression in macula densa cells gradually reappeared and again achieved control levels by *day 20* post partum. Breton et al. observed that both the OT receptor mRNA levels as well as the OT binding capacity of rat renal extracts were significantly lower at term. Although in the rat kidney there was no indication for another than the uterine-type OT receptor gene expression, the mechanisms controlling the expression of OT receptor genes in uterus, pituitary, and kidney are obviously different.

OT injected intraperitoneally caused dose-dependent increases in urinary osmolality, natriuresis, and kaliuresis in the rat. Moreover, the injection of OT evoked concomitant release of ANP and natriuresis. Soares et al. recently provided evidence for the hypothesis that OT acts in concert with ANP to induce natriuresis and kaliuresis via the common mediator cGMP. In rats, the OT-induced natriuresis was caused mainly by decreased tubular Na^+reabsorption and was accompanied by increases in both urinary cGMP and NO_3^- excretion. Although the natriuretic action of ANP is mediated by activation of renal guanylyl cyclase A (GC_A) receptors localised in glomeruli, their afferent and efferent arterioles, and the tubules, OT is hypothesized to induce natriuresis by activation of renal NO synthase localised in macula densa cells. Presumably, binding of OT to their receptors on NOergic cells in the proximal tubules and macula densa leads to an increase of intracellular [Ca^{2+}], followed by Ca^{2+}/calmodulin-induced stimulation of NO synthase and activation of the soluble guanylate cyclase by NO. The consequent release of cGMP could then mediate natriuresis and kaliuresis via closure of Na^+ and K^+ channels. In cortical collecting duct cells, it has been shown that the NO-inhibited Na^+ transport is associated with increased cGMP content. Overall in rats, OT as well as ANP induce a concomitant reduction of both extracellular and intracellular fluid volume in states of increased body fluid volume.

The signal transduction of the renal OT receptor has been evaluated in various kidney epithelial cells in culture, e.g., LLC-PK_1. In these cells, OT induced stimulated phosphoinositide hydrolysis and transiently increased cytosolic [Ca^{2+}]. OT was also able to stimulate the soluble guanylate cyclase and increased the intracellular level of cGMP in LLC-PK_1 cells. Moreover, OT was found to stimulate the synthesis of prostaglandins in a dose-dependent manner in kidney homogenates. What is known about the OT effects on kidney function in species other than rats? In rabbits, OT was shown to affect apical sodium conductance of the microperfused cortical collecting duct through OT receptors distinct from the vasopressin receptors. The effects were inhibited by the addition of an OT antagonist. Although the natriuretic response to OT has also been described for conscious dogs, it probably does not occur in humans and nonhuman primates. Thus the contribution of OT to renal physiology in primates including humans, if any, remains uncertain.

Heart and Cardiovascular System E

Peripherally injected OT decreases mean arterial pressure in rats. Even in the absence of a central control mechanism, OT is able to reduce the heart rate and the force of atrial contractions in isolated atria from perfused rat hearts. An OT antagonist reversed the bradycardia caused by OT. Moreover, administration of OT at high concentrations (1 ¼M) leads to the stimulation of ANP release. An OT antagonist first inhibited this ANP release, and after prolonged perfusion, it finally decreased the OT-induced ANP release below that of control hearts. This suggested that intracardial OT stimulates ANP release in the heart. Gutkowska et al. proposed that OT and ANP act in concert in the control of body fluid and in cardiovascular homeostasis. In favour of this hypothesis, OT receptor transcripts and OT binding sites were shown to be present on atrial and ventricular sections as detected by in situ hybridisation and autoradiograpy, respectively. Furthermore, the OT receptor gene is expressed in all chambers of the rat heart, and the analysis of the RT-PCR products indicated the presence of the uterine-type OT receptors in the heart. Although the OT receptor mRNA levels in the atria were found to be higher than in the ventricles, overall the OT receptor mRNA levels were calculated to be at least 10 times lower than the OT receptor mRNA level present in the uterus of a nonpregnant rat.

The rat heart was also shown to be a site of OT synthesis. OT was detected in the effluent of isolated heart perfusates as well as in the medium of cultured atrial myocytes. OT concentrations were found to be higher in the atria than in the ventricles. In the right atrium, OT concentrations ~20-fold higher than those in the rat uterus have been measured. In contrast, the OT mRNA level in the heart tissues was found to be lower than that in the rat uterus. This discrepancy argues against the postulated abundant biosynthesis of cardiac OT. The relatively low concentrations of OT in the heart chambers and the high OT doses required for ANP release would be more compatible with paracrine or autocrine effects of cardiac OT, particularly in the right atrium. Although the OT quantities released from the perfused heart are not sufficient to substantially change the plasma concentrations of OT, they may contribute to the natriuretic action via stimulation of ANP release. Presumably, blood volume expansion via baroreceptor input to the brain causes the release of OT that circulates to the heart. OT-induced ANP release in the heart may be achieved after activation of OT receptors and subsequent elevation of intracellular [Ca^{2+}], which in turn could stimulate exocytosis and ANP secretion. ANP then exerts a negative chrono- and inotropic effect via activation of guanylyl cyclase and release of cGMP. Finally, a rapid reduction in the effective circulating blood volume is produced by an acute reduction in cardiac output, coupled with ANP's peripheral vasodilating actions. The ANP released would also act on the kidneys to cause natriuresis, and ANP acts within the brain to inhibit water and salt intake, leading to a gradual recovery of circulating blood volume to normal. Since the plasma concentration of both OT and ANP were found to be increased after parturition, the OT-stimulated ANP release might be at least partly responsible for the massive diuresis observed postpartum.

The effects of repeated subcutaneous OT injections on blood pressure and heart rate were investigated in spontaneously hypertensive rats. Surprisingly, when OT were given for 5 days, a sustained decrease in blood pressure was observed in male but not female rats, whereas the heart rate was unaffected. Moreover, acute versus chronic OT treatments caused opposite effects on blood pressure, and these effects were modified by female sex hormones.

It appears that the complete OT system is present in the vasculature of the rat. The OT concentrations in the aorta and vena cava were reported to be even higher than those in the right atrium of the heart. Thus OT might play a direct role in volume and pressure regulation in a paracrine/autocrine manner. In the case of the umbilical-placental vasculature, OT was reported to be a far more potent vasoconstrictor than vasopressin. OT was therefore proposed to be involved in the closure of the umbilical vessels at birth. A uterine-type OT receptor could also mediate vasodilatory responses in human vascular endothelical cells.

OT-induced cardiac effects were also reported for some other species. The cardiac effects of neurohypophysial hormones were studied in the frog, *Rana tigrina*, and in the snake, *Ptyas mucosa*. In both species, OT produced dose-related cardiac effects in isolated atrial preparations. In the frog, vasotocin was the most potent hormone, whereas in the snake, OT produced the most potent dose-related positive inotropic changes. Furthermore, OT was found to modulate the intrathoracic sympathetic ganglionic neurons regulating the canine heart. When OT was injected into right atrial ganglionated plexi, heart rate and atrial forces were reduced in 5 of 10 dogs studied. However, no cardiac changes occurred when OT was injected into left atrial or ventricular ganglionated plexi. In this study, the connectivity with the central nervous system (CNS) was found to be necessary to elicit consistent OT-induced responses. Although OT elicited neuronal and concomitant cardiovascular responses in most dogs when administered adjacent to spontaneously active intrinsic cardiac neurons, OT injection into the superior vena cava only evoked a slight systemic hypotension in two of seven dogs.

Other Localisations F

Thymus 1: The thymus is the primary lymphoid organ reponsible for the selection of the peripheral T-cell repertoire. Neuropeptides and their receptors have been described in the thymus, supporting the concept of a close dialogue between the neuroendocrine and the immune systems at the level of early T-cell differentiation. The neurohypophysial peptides have been shown to trigger thymocyte proliferation and could induce immune tolerance of this conserved neuroendocrine family. OT has been identified in human thymus by immunoreactivity and at the transcriptional level. OT was present in the thymus in surprisingly large amounts and was found to be restricted to certain subtypes of thymic epithelial cells. Due to its higher expression level in the thymic epithelium, OT was proposed as the self-antigen of the neurohypophysial family. OT was found to be colocalised with the cytokeratin network of thymic epithelial cells and not within secretory granules. The peptide is therefore not secreted but behaves like an antigen presented at the outer surface of the cell. Thymic OT also behaves as a cryptocrine signal targeted at the epithelium membrane from where it is able to interact with neurohypophysial peptide receptors expressed by pre-T cells. OT receptors are predominantly expressed by cytotoxic $CD8^+$lymphocytes, and they transduce signals via the phosphoinositide pathway. In pre-T cells, OT was found to induce the phosphorylation of focal adhesion kinase. Thus it was suggested that OT actively intervenes in the program of T-cell differentiation both as a neuroendocrine self-antigen and as a promoter of T-cell focal adhesion following a cryptocrine pathway. OT receptors detected in thymic membranes or on thymocytes revealed a ligand selectivity similar to that of uterine OT receptors. Caldwell et al. reported that sexual activity leads to decreases of OT receptor densities in the thymus of rats. In human thymus extracts, the content of the OT immunoreactivity declined with increasing age, whereas in rat thymic extracts, it was reported to increase during aging. The age-dependent changes might be linked to thymic involution. Some immune pathologies in humans may be explained by thymic OT being involved in T-cell-positive selection and activation.

Fat Cells 2: In adipocytes, OT has a so-called insulin-like activity in that it stimulates glucose oxidation and lipogenesis and increases pyruvate dehydrogenase activity. Treatment of rat adipocytes with lipolytic stimuli such as glucagon or isoprenaline increased the conversion of choline into phosphatidylcholine, and this lipolytic effect was antagonised by OT. In rat adipocytes, OT at a concentration of 1 nM markedly increased phosphoinositide breakdown. OT also elicited a transient increase in intracellular [Ca^{2+}] and stimulated the protein kinase C activity. Moreover, human fat cells possess a plasma membrane-bound H_2O_2 generating system that is sensitive to extracellular stimuli. OT as well as insulin were able to activate this system. It was suggested that the H_2O_2 produced might participate in the regulation of fat cell differentiation and/or maintenance of the differentiated state. Populations of low- and high-affinity OT receptors have been described in rat adipocytes.

Pancreas 3: OT and AVP have been identified in human and rat pancreatic extracts at higher concentrations than those found in the peripheral plasma. However, a local synthesis of these peptides within this organ has not yet been established. According to most studies in several species, the neurohypophysial hormones induce the release of glucagon and, to a lesser extent, insulin from the pancreas. During in situ perfusion of the rat pancreas with OT, a marked stimulation of glucagon release and a modest stimulation of insulin release were observed. With the use of isolated islets from rat pancreas, OT elicited a stimulation of glucagon release but failed to influence insulin release in a culture medium with low glucose content. The glucagonotrophic action of OT was greatly diminished in the presence of a higher concentration of glucose. Similar findings were reported in studies with conscious dogs. With the use of the microdialysis technique, OT administered directly in the pancreas of the rat stimulates the release of both insulin and glucagon. OT binding sites were identified in the periphery of the islets of Langerhans, corresponding to the localisation of the glucagon-producing α-cells. In isolated mouse islets, OT produced an increase in somatostatin, glucagon, and insulin release, and it amplified the glucose-induced insulin release probably via stimulation of the phosphoinositide metabolism. In humans, OT evoked a rapid surge in plasma glucose and glucagon levels followed by a later increase in plasma insulin and epinephrine levels. The effects of OT on plasma glucagon and epinephrine levels were potentiated by hypoglycemia. OT was also found to potentiate glucose-induced insulin secretion. In contrast, Page et al. observed no effects of OT on the decline or recovery of blood glucose concentrations or on the plasma glucagon response to insulin-induced hypoglycemia in humans.

Impaired glucose tolerance and hyperinsulinemia are common features of obesity. Interestingly, the plasma levels of OT were found to be fourfold higher in male and female obese subjects compared with control subjects. OT rises in hypoglycemia, and this response is partially inhibited by dexamethasone. The OT rise in response to insulin-induced hypoglycemia was reduced in obese men. Pretreatment with the opioid antagonist naloxone enhanced the OT response to hypoglycemia in obese men and suggested an abnormal activity of endogenous opioids in obesity. In women, unlike men, endogenous opioids did not modulate OT release during insulin-induced hypoglycemia.

A pancreatic OT receptor was cloned from the rat insulinoma cell line RINm5F. CHO cells that have been transfected with this receptor responded to OT with an increase in cytosolic Ca^{2+} concentration and inositol phosphate levels as well as arachidonic acid release and PGE_2 synthesis. Amino acid sequence homology, binding specificity, and the ligand-activated signal transduction pathways suggested that the pancreatic OT receptors expressed in RINm5F cells are indistinguishable from the uterine-type OT receptors that have been described in other tissues.

Adrenal Gland 4: Ang and Jenkins first identified immunoreactive OT and AVP in human and rat adrenal glands. OT predominanted over AVP, and both peptides occurred at concentrations far greater than those found in plasma. Immunohistochemical studies in adrenal glands from rat, cow, hamster, and guinea pig showed that OT was localised in both the cortex and medulla in all these species. In the cortex, the OT immunoreactivity was most intense in the zona glomerulosa, whereas in the medulla, the OT staining was higher but revealed more species variation. Bovine adrenal medulla membranes revealed high-affinity low-capacity OT binding sites. OT binding was also detected on primary monolayer cultures of bovine adrenal chromaffin cells. However, the capacity of these sites was near the detection limit, and OT concentrations in the micromolar range were necessary to stimulate the phosphoinositide pathway. From the few studies that have addressed the role of OT in the adrenal gland, no clear picture has emerged. Perfusion of the isolated rat adrenal gland with OT at 100 nM inhibited the acetylcholine-stimulated aldosterone secretion, whereas smaller doses of OT given as a bolus were able to stimulate aldosterone secretion in the intact perfused rat adrenal gland, but not in superfused adrenal cells. Legros et al. hypothesized that OT acts also at the adrenal gland level to decrease cortisol release and/or synthesis in humans. In addition, a proliferative effect of OT was suggested by the observation that the number of chromaffin cells in the adrenal medulla of rats was increased in response to OT.

The list of peripheral tissues containing OT receptors is certainly not complete. Recently, functional OT receptors have also been discovered in primary cultures of human osteoblasts and in a human epithelial osteosarcoma cell line (Saos-2). Further studies are required to show whether OT is a bone anabolic agent as suggested by the authors.

The Central Oxytocin System

Localisation Profile A

Localisation of OT 1: In the CNS, the OT gene is primarily expressed in magnocellular neurons in the hypothalamic PVN and SON. Action potentials in these neurosecretory cells trigger the release of OT from their axon terminals in the neurohypophysis. In the PVN, two populations of OT-staining neurons have been identified: magnocellular neurons that terminate in the neurohypophysis and parvocellular neurons that terminate elsewhere in the CNS. Only a small fraction (0.2%) of the OT neurons were calculated to possess axon collaterals to both the neurohypophysis and the extrahypothalamic areas. Although few OT perikarya were observed other than in the magnocellular nuclei, OT fibres and endings have been described in various brain areas in the rat: the dorsomedial hypothalamic nucleus, several thalamic nuclei, the dorsal and ventral hippocampus, subiculum, entorhinal cortex, medial and lateral septal nuclei, amygdala, olfactory bulbs, mesencephalic central gray nucleus, substantia nigra, locus coeruleus, raphe nucleus, the nucleus of the solitary tract, and the dorsal motor nucleus of the vagus nerve. OT fibres also run toward the pineal gland and the cerebellum, with most of them continuing toward the spinal cord. A few OT fibres end on the portal capillaries in the median eminence.

OT concentrations in the extracellular fluid of the SON were calculated to be >100- to 1,000 fold higher than the basal OT concentration in plasma, i.e., more than 1–10 nM. High-frequency electrical discharges of OT neurons as they occur, e.g., during the milk ejection reflex, might release even higher local OT concentrations.

Plasma OT does not readily cross the blood-brain barrier, and there is no relationship between the release of OT into the blood by the neurohypophysis and the variations in OT concentrations in the cerebrospinal fluid (CSF). Peripheral stimulations such as suckling or vaginal dilation that elicit large increases in plasma OT may or may not change the concentration of OT in the CSF. As shown in rats, electrical stimulation of the neurohypophysis only evokes the release of OT into the blood, whereas stimulation of the PVN elicits a release of OT into the blood and into the CSF. After hypophysectomy, OT disappears from the blood, whereas its concentration increases in the CSF. The OT in the CSF is probably derived from neurons that extend to the third ventricle, the limbic system, the brain stem, and the spinal cord. In the CSF, OT is normally present at concentrations of 10–50 pM, and its half-life is much longer (28 min.) than in the blood (1–2 min.). In humans and in monkeys, a circadian rhythm in the OT concentrations in the CSF has been found with peak values at midday. No such circadian rhythms have been observed in the CSF of rats, cats, guinea pigs, or goats. Circadian rhythms have never been observed in plasma OT concentrations.

Further complicating factors are the presence of an OT-like peptide and the appearance of conversion products of OT that possibly exert their effects exclusively on the brain, e.g., to influence learning and memory processes. Moreover, OT fragments such as OT-(1-6) or OT-(7-9) could cross the blood-brain barrier more easily.

Localisation of OT Receptors 2: Experiments with primary cell cultures showed that OT receptors are localised both on hypothalamic neurons and astrocytes. Concerning the regional distribution of OT binding sites in the brain, marked species differences have been observed. In rats, OT receptors are abundantly present in several brain regions, i.e., some cortical areas, the olfactory system, the basal ganglia, the limbic system, the thalamus, the hypothalamus, the brain stem, and the spinal cord. In the adult rat, a high density of OT receptors is found in the dorsal peduncular cortex, the anterior olfactory nucleus, the islands of Calleja and ventral pallidum cell groups, the limbic system (bed nucleus of the stria terminalis, central amygdaloid nucleus, ventral subiculum), and the hypothalamic ventromedial nucleus. OT receptor mRNA was detected in brain areas mostly coinciding with the occurrence of OT binding sites. OT receptors were detectable in all spinal segments, but in low amounts and restricted to the superficial layers of the dorsal horn. No major differences in receptor distribution were observed between male and female brains. Notably, the distribution pattern of OT binding sites is markedly different from that of binding sites for AVP. Whenever present in the same area, OT and AVP binding sites were located in different parts of the area.

The distribution and numbers of OT binding sites undergo major changes during development. As shown by Tribollet et al., only a fraction of OT receptors is constantly present during the development. Some OT receptors are only transiently expressed on neurons, e.g., during infancy or during maturation. In male and in female rats, two critical periods during the development were recognised: the third postnatal week, which precedes weaning, and puberty. For example, in the infant brain, the cingulate and retrosplenial cortex as well as the substantia gelatinosa of the spinal cord contained the highest densities of OT binding sites, whereas low or undetectable numbers of OT receptors were observed in these areas in the adult rat brain. Conversely, OT receptors in some other areas were abundant in the adult brain but undetectable before puberty, e.g., in the olfactory tubercle (Calleja islands and ventral pallidum cell groups) and in the hypothalamic ventromedial nucleus. During aging, however, the number of OT binding sites decreased again in the latter areas. This was probably mediated by the markedly lower level of plasma testosterone in aged rats. In fact, the expression of OT receptors in the olfactory tubercle and in the ventromedial hypothalamic nucleus was shown to be dependent on gonadal steroids, and testosterone treatment of aging rats could restore normal adult levels of OT receptors in these areas.

As already mentioned, there is a high diversity of OT receptor distribution between different species. For example, the ventral subiculum in the hippocampus contains high densities of OT binding sites in the rat, whereas in the guinea pig, the hamster, the rabbit, and the marmourset, no OT binding sites were detected in this area. In the rabbit, no OT receptors were found in the hypothalamic ventromedial nucleus. In monogamous versus polygamous voles, OT receptor distribution was shown to reflect social organisation. In human brains, dense OT receptor binding sites were visualised in the pars compacta of the substantia nigra unlike in several other species examined so far. Thus, in humans, nigrostriatal dopamine neurons could be a target for OT, and the OT system may be involved in motor and other basal ganglia-related functions. Strong OT binding intensity was also observed in the basal nucleus of Meynert, but OT binding sites were lacking in the hippocampus, amygdala, entorhinal cortex, and olfactory bulb of human brains.

In a few brain areas such as the ventral hippocampus and the bed nucleus of the stria terminalis (BNST), the reactivity of neurons to OT could be related with OT axon endings and OT receptors. The distribution of OT binding sites in the rat spinal cord was shown to coincide with that of the OT innervation, suggesting that OT is involved in sensory and autonomic functions. In most other brain areas, it was not possible to identify a clear relationship between the presence of the whole OT system and the physiological data. So, in male rats, Hosono et al. identified a functional OT receptor system in the subfornical organ, where in binding studies the expression is almost neglected. Marked receptor-peptide mismatches exist in some brain regions, for example, in the amygdala, where very few OT afferents but highest OT binding activities have been found. On the other hand, one has to recognise that OT binding sites in regions with high OT concentrations may be downregulated to a level that may not be detectable by autoradiography using radioligands. In the lactating rat, for instance, OT binding sites have been found to be nearly undetectable in den PVN and SON. As soon as 5–20 min. after intracerebroventricular injection of an OT antagonist, a strong increase of OT binding sites was noticed in the magnocellular nuclei. In this case, the presence of the OT antagonist may prevent the downregulation of OT receptors induced by the high concentrations of OT that are released into the magnocellular nuclei during lactation.

In the brain, estrogen has only a modest influence on the synthesis of OT, but it has a pronounced effect on the regulation of the OT receptor. OT receptors (but not AVP receptors) are regulated by gonadal steroids in the rat brain in a complex fashion. Castration and inhibition of aromatase activity reduced, whereas estradiol and testosterone increased OT binding, particularly in regions of the brain assumed to be involved in reproductive functions, such as the ventrolateral part of the hypothalamic ventromedial nucleus (VMN) and the islands of Calleja and neighbouring cell groups. Estrogen treatment increased the affinity of OT receptors in the medial preoptic area-anterior hypothalamus and increased both the density and the area of OT binding in the rat VMN. Progesterone caused a further increase in receptor binding and was required for a maximal extension of the area covered by OT receptors. In another study, chronic progesterone treatment increased basal OT receptor density in the limbic structures, decreased it in the ventromedial nucleus, and prevented estrogen-induced increases in ligand binding in all areas studied with the exception of the medial preoptic area. Glucocorticoids have also been reported to modulate cerebral OT receptors. The brain OT receptor varies across species not only in its distribution but also in its regional regulation by gonadal steroids. For example, estrogen increased OT receptor binding in the rat brain but reduced OT receptor binding in the homologous regions of the mouse brain.

Hypothalamus-Neurohypophysis B: The magnocellular neurons also release OT and VP from their perikarya, dendrites, and/or axon collaterals. Although the amount of release is small compared with the amount released from the neurohypophysis, the concentration of OT and VP in the extracellular fluid of the SON resulting from this somatodendritic release has been calculated to be 100- to 1,000-fold higher than the basal plasma concentration. Intranuclear release of these peptides occurs in response to a wide variety of stimuli, including suckling; parturation; hemorrhage; certain kinds of stress such as fever, physical restraint, and pain; mating and territorial marking behaviours; dehydration; administration of hypertonic solutions; and a range of pharmacological stimuli. The magnocellular neurons display characteristic activity patterns that are associated with particular secretion patterns for the peptide. For example, OT neurons respond to such stimuli as hyperosmolarity with small increases in spontaneous firing rate, whereas during lactation the same cells display explosive synchronised bursts of activity associated with a pulsatile release of OT into the circulation to cause contraction of mammary smooth muscle and milk let-down. PVN neurons can be identified as either AVP or OT secreting on the basis of their spontaneous discharge patterns. Interestingly, in most cases, central release patterns of OT are accompanied by peripheral ones, whereas release of AVP is not. In the rat, OT is released into the plasma without AVP, e.g., by suckling, parturition, stress, and nausea.

OT is required for successfull milk ejection in response to suckling as confirmed by the phenotype of OT knock-out mice. All offspring of these mice died shortly after birth because of the dam's inability to nurse. Postpartum injections of OT to the OT-deficient mothers restored milk ejection and rescued the offspring. OT released into the SON probably plays an essential role in the milk-ejection reflex, facilitating suckling-induced electrical activation of OT neurons. So, the injection of an OT antagonist into the SON prevented the OT-induced facilitation and completely interrupted the milk-ejection reflex. During parturition and suckling, OT is released within the SON and apparently excites via a short positive-feedback loop the same cells by which it is produced and secreted. This autoexcitatory mechanism leads to further amplification of local and/or neurohypophysial OT release and ensures the synchrony of firing activity among oxytocinergic neurons. The underlying molecular mechanism of this putative autoexcitation is unclear. It was suggested that OT depresses the synaptic GABAergic input from perinuclear interneurons projecting into the SON. In addition to the magnocellular nuclei, the BNST may participate in the control of neuroendocrine responses during lactation. Injection of OT into the BNST increases the frequency of milk ejections, and electrophysiological recordings showed increased activity of BNST neurons coincident with this facilitatory effect.

All the physiological situations during which large amounts of OT are released into the blood are characterised by ultrastructural changes in the magnocellular nuclei, e.g., reduced astrocytic coverage of oxytocinergic somata and dendrites, increases in GABAergic synapses, increases in the juxtaposition of the membranes of the perikarya and of the dendrites between adjacents neurons, and increases in the contact area between neurosecretory terminals and the perivascular space. These morphological changes are reversible with cessation of stimulation, affect exclusively oxytocinergic neurons, and may serve to facilitate and maintain the characteristic synchronised electrical activity of these neurons at milk ejection . Neuronal network rearrangement may also occur after behavioural experience. So, in maternally experienced ewes, parturition-induced increases in the expression of OT receptor mRNA in the PVN and the islands of Calleja were potentiated.

Adenohypophysis C

It has been suggested that some hypothalamic OT reaches the anterior pituitary lobe via the hypothalamo-pituitary portal vasculature. OT might thus be able to influence anterior pituitary hormones as a hypothalamic regulating factor. OT is present in nerve terminals in the median eminence. It was found to be released into the portal vessels, and specific OT receptors are present in the rat adenohypopophysis. Another pathway for OT delivery to the adenohypophysis might be the short portal vessels connecting the posterior and anterior lobes. OT may participate in the physiological regulation of the adenohypophysial hormones prolactin, ACTH, and the gonadotropins.

There was a long controversy on whether OT released in response to suckling was responsible for the concomitant secretion of prolactin from the adenohypophysis. During suckling and under stress, both hormones are quantitatively predominant among the factors released. OT could only act as prolactin releasing factor when the dopamine levels are low, e.g., during the brief periods of dopamine withdrawal that characterise the onset of prolactin secretion under various physiological stimuli. With the use of a cell-specific targeting approach, it was shown that a subpopulation of lactotrophs respond to OT. Breton et al. demonstrated that the pituitary OT receptor gene expression is restricted to lactotrophs and dramatically increases at the end of gestation or after estrogen treatment. These findings question a direct function of OT on pituitary cells other than lactotrophs and suggest that OT might exert its full potential as a physiological prolactin-releasing factor only toward the end of gestation.

The major endocrine response to stress is via activation of the hypothalamic-pituitary-adrenal axis. ACTH secretion from the anterior pituitary is primarily regulated by CRH and AVP synthesized in neurons of the PVN. Unlike ACTH, plasma OT does not increase in response to all kinds of stress. In rats, OT can potentiate the release of ACTH induced by CRH. OT was also reported to stimulate ACTH secretion from corticotrophs in fetal and adult sheep. CRH is responsible for the immediate secretion of ACTH in response to stress. However, when the CRH levels are decreased following prolonged stress, the persistent level of OT in the median eminence could become important for the delayed ACTH response and for the generation of pulsatile ACTH secretory bursts.

In contrast, OT infusion into human volunteers actually inhibited the plasma ACTH responses to CRH. Suckling and breast stimulation in humans produced an increase in plasma OT and a decrease in plasma ACTH level. The observed negative correlation of both hormones indicates an inhibitory influence of OT on ACTH/cortisol secretion under a certain physiological condition in humans. Conclusively, OT might control ACTH release under some physiological conditions in a species-specific manner.

Gonadotropes in the adenohypophysis synthesize and secrete the two gonadotropin hormones, luteinizing hormone (LH) and follicle-stimulating hormone (FSH). Although gonadotropin-releasing hormone (GnRH) is believed to be the primary secretagogue for LH, OT has also been shown to stimulate LH release. OT administered to proestrous rats caused advancement of the LH surge and earlier ovulation. In addition, OT antagonists inhibited the peak of LH at proestrous. On the other hand, OT receptors have not been detected on gonadotropes but were identified on a gonadotrope-derived cell line. Indirect effects of OT on LH release have also been discussed. OT has been observed to synergistically enhance GnRH-stimulated LH release. OT may sensitise the pituitary before full GnRH stimulation. In human females, preovulatory OT administration promoted the onset of the mid-cycle LH surge. Overall, the physiological connection between OT and LH release has yet to be definitively established.

Centrally Mediated Autonomic and Somatic Effects D.

Cardiovascular Regulation 1: In the rat, OT-containing axons terminate in several brain stem nuclei known to be involved in cardiovascular control. High concentrations of OT are found in the nucleus of the solitary tract, which receives sensory inputs from the viscera, including baroreceptors of the cardiovascular system. Nevertheless, microinjection of OT into the nucleus tractus solitarius did not change the mean arterial pressure or the heart rate in the rat. In contrast, injection of OT into the dorsal motor nucleus of the vagus reduced the heart rate, and this effect was eliminated by an OT antagonist. In humans and rats, the bolus intravenous administration of OT is often associated with a decrease in blood pressure.

A role of central OT with respect to cardiovascular control has been established in many studies. Central pretreatment of rats with an OT antisense oligodeoxynucleotide attenuated the mean arterial pressure and heart rate responses induced by substance P. This suggests that OT neurons in the PVN mediate the increases in blood pressure and heart rate induced by stimulation of substance P receptors in the forebrain. An OT antisense oligonucleotide injected into the PVN abolished the tachycardia produced by shaker stress in rats, indicating that OT may act as mediator of stress-induced tachycardia. In addition, neuropeptide Y and peptide YY, which are central and peripheral modulators of cardiovascular and neuroendocrine functions, triggered OT release from perfused isolated terminals of the rat neurohypophysis. There is further evidence for an interaction between the central oxytocinergic and vasopressinergic systems in cardiovascular control. So, central injection of AVP (1–10 pmol) increased the mean arterial pressure and heart rate, and both responses were found to be enhanced in rats pretreated with OT. Recent findings suggested a putative interaction of OT with the ANPergic system. Although atrial stretch releases ANP from cardiac myocytes, the response to acute blood volume expansion was found to be markedly reduced after elimination of neural control. Volume expansion distends baroreceptors in the right atria, carotid-aortic sinuses, and kidney and in turn alters afferent input to the brain stem and the hypothalamus. Because axons of the ANP neurons contained only low amounts of ANP (1,000 times less than in right atrium of the heart), it was hypothesized that hypothalamic ANP neurons cause release of OT, which circulates to the right atrium to stimulate ANP release. Because ANP has a negative ino- and chronotropic effect in the atrium, this would produce an acute reduction in cardiac output and, coupled with peripheral vasodilating actions, causes a rapid reduction in effective circulating blood volume.

Analgesia 2: Analgesic effects of OT have been reported in most but not all studies in mice, rats, dogs, and humans. Certain stimulations such as vaginal dilation led not only to a rise in plasma OT concentrations but also to an increase of the pain threshold. Dogs with neck and back pain caused by spinal cord compression had significantly more OT in their CSF than clinically normal dogs. Analgesia was observed in rats after injecting OT into the lateral ventricles. On the other hand, systemic OT did not produce analgesia in rats. Thus it was concluded that the observed increase in response latency in the hot-plate test may result from the sedative and vasoconstrictive effects of OT. Because the OT antagonist did not alter response latency on the hot-plate test, it seems unclear whether endogenous OT exerts a tonic effect on the pain threshold in rats.

In humans, intrathecal injection of OT was effective in treating low back pain for up to 5 h. An OT antagonist and the opiate receptor-blocker naloxone could reverse OT-induced analgesia. OT also increased β-endorphin and L-encephalin contents in the spinal cord, whereas an OT antagonist caused a decrease in the concentration of these opioids. Moreover, OT levels were elevated in the CSF of patients with chronic low back pain, perhaps a compensatory response to the painful condition. In a clinical case study, a high concentration (300 ¼g) of

OT injected intravenously was reported to evoke strong analgesic effects lasting more than 70 min. in a patient with intractable cancer pain at a time when opiates were no longer effective.

Motor Activity 3: Centrally administered OT can induce or modify several forms of behaviour together with the associated motor sequences. OT increased general motor activity, and OT antisera decreased this hyperactivity and seizures in a complementary fashion. In this context, OT may possibly act at the spinal level. OT changed the spontaneous motor activity in female rats in strong dependence on steroid hormones. Treatment with low OT doses led to a decrease in peripheral locomotor activity, whereas increasing doses of OT provoked sedative effects as indicated by a suppression of locomotor activity and rearing. A maximal effect was obtained within 1 h and thereafter, the behaviour gradually returned to normal within 24 h. During static muscle contraction, blood pressure and heart rate reflexly increase as shown in anaesthetised cats. In this respect, OT may participate to regulate cardiovascular responses elicited by contraction of skeletal muscle.

Thermoregulation 4: In contrast to centrally administered AVP, for which antipyretic actions have been well documented, OT evoked, if at all, mostly weak antipyretic effects at higher concentrations. In rabbits, intracerebroventricular administered OT produced small but long-lasting hyperthermias that did not exceed 0.4°C. Intracisternally injected OT to adult male mice significantly increased colonic temperatures, with the maximal rise in temperature occurring 30 min. after administration of the peptide. OT also antagonised the hypothermia produced by other peptides such as bombesin or neurotensin. Conversely, whenever the body temperature was elevated in urethan-anaesthetised rats, e.g., in response to prostaglandin fever, OT was released into the ventral septal area, but not into the hippocampus. OT could also modulate the antipyretic action of AVP in rats. Possibly, the effects of OT with respect to thermoregulation may be physiologically significant during parturition and lactation.

Gastric Motility 5: In conjunction with the natriuretic effects of circulating OT, Verbalis et al. suggested the concept of a coordinated central and peripheral OT secretion as a mechanism for regulating body solute homeostasis in rats. The following findings support this concept. OT inhibits food intake in fasted rats and NaCl intake in hypovolemic rats. Inhibition of food and/or NaCl intake stimulates expression of c-*fos* in parvocellular and magnocellular OT neurons and indicates simultaneous activation of centrally projecting and pituitary-projecting OT neurons. Selective inactivation of brain cells containing OT-receptive elements leads to a disinhibition of NaCl intake similar to that produced by OT antagonists. In particular, gastric motility is inhibited by microinjection of OT into the dorsal motor nucleus or by intracerebroventricular administration of OT in rats. An OT antagonist administered intracerebroventricularly alone caused an increase in baseline gastric motility. In addition, stimulation of gastric vagal afferents by systemic administration of cholecystokinin (CCK) inhibited gastric motility, reduced food intake, and stimulated pituitary secretion of OT in rats. McCann and Rogers provided evidence that central OTergic neurons influence gastric motility and secretion by increasing the excitability of brain stem vagal neurons, i.e., sensory neurons in the nucleus tractus solitarius and motor neurons in the dorsal motor nucleus, which are both related to gastric function. According to Esplugues et al., the inhibitory effect of OT on the distension-stimulated gastric acid secretion in the rat perfused stomach is mediated by a nervous reflex involving a neuronal pathway that includes NO synthesis in the dorsal motor nucleus of the vagus.

For OT to have physiological relevance for maintenance of solute homeostasis, a coactivation of magnocellular and parvocellular OT ergic systems has been postulated: central, i.e., parvocellular OT as inhibitor of solute ingestion, and peripheral, i.e., pituitary secreted OT to enhance solute excretion via renal sodium excretion. However, pituitary OT secretion from magnocellular neurons is not always linked to decreased gastric motility in rats. For example, nipple attachment and sucking by pups, a well-known stimulus for neurohypophysial secretion of OT, did not decrease gastric motility in lactating rats. In view of the fact that OT promotes reproductive behaviours, it appears quite plausible that OT simultaneously acts as inhibitor for food intake, since both behavioural responses would conflict with each other. However, some findings do not fully support this concept or need to be clarified. First, central administration of an OT antagonist did not block the effects of CCK and LiCl to inhibit gastric motility. This suggests that the parvocellular OT projections modulate the activity of intrinsic brain stem reflex arcs and do not exert a direct control over vagal efferent outputs. Second, OT receptors and mRNA localisations have been reported to occur predominantly on vagal neurons of the dorsal motor nucleus, which are generally excitatory to gastric motility, whereas OT is known to inhibit motility via neurons of the nucleus tractus solitarius. Finally, marked interspecies differences have been reported with respect to OT's relevance for maintenance of solute homeostasis. Whereas in rats nausea-producing agents and CCK each stimulated OT rather than AVP release, in humans and monkeys vasopressin but not OT secretion was provoked in response to these agents. In humans, nausea is a powerful stimulator of VP release induced by centrally acting emetics such as apomorphine, whereas OT is only weakly stimulated by apomorphine. In the rat, the converse is true. OT is stimulated by emetic agents, and VP shows little change. Therefore, we have little evidence that OT is involved in the physiology of emesis in humans. Intracisternal injection of OT also failed to induce emesis in dogs.

Hydromineral Regulation and Osmoregulation 6

In rats, the concerted release of OT and AVP plays a key role in osmoregulation through the effects of natriuresis and diuresis. Osmoreceptors are located in systemic viscera and in central structures that lack the blood-brain barrier. The nucleus tractus solitarius and lateral parabrachial nucleus receive neural signals from baroreceptors and are responsible for inhibiting the ingestion of fluids under conditions of increased volume and pressure and for stimulating thirst under conditions of hypovolemia and hypotension. Osmosensitive neurons combine with endogenously generated osmoreceptor potentials to modulate the firing rate of magnocellular neurons. In magnocellular neurons isolated from the SON of the adult rat, stretch-inactivated cationic channels transduce osmotically evoked changes in cell volume into functionally relevant changes in membrane potential. As judged from the distribution of Fos immunoreactivity, a fraction of oxytocinergic neurons in the subnuclei of the PVN were found to be activated during the satiety process of sodium appetite. Blackburn et al. provided evidence that salt appetite is regulated by both sodium- and osmolality-sensing mechanisms in rats. Using a selective receptor inactivation approach, these authors concluded that in contrast to the central ANP receptors that mediate both sodium- and osmolality-induced inhibition of NaCl ingestion, the central OT receptors primarily mediate osmolality-induced inhibition of NaCl ingestion.

In anaesthetised rats, electrical stimulation of renal afferents provided evidence that sensory information originating in renal receptors changes the activity of oxytocinergic neurons in the PVN. Thus renal receptors could contribute to the hypothalamic control of AVP and OT release into the circulation. In humans, there is neither an OT response to changes in plasma osmolality nor is there a natriuretic response to large doses of OT.

Central Behavioural Effects

Sexual Behaviour A

The "classical" peripheral target tissues for OT, uterus and mammary glands, are both organs linked to reproduction. There is also evidence that OT homologs in fish, amphibians, and reptiles as well as in molluscs and annelids are important in the control of reproductive behaviours. In the mollusc *Lymnaea stagnalis*, the conopressin gene encoding the putative ancestral receptor to the VP/OT receptor family is expressed in neurons that control male sexual behaviour, and its gene products are present in the penis nerve and the vas deferens. Thus a phylogenetically old connection may exist between systems controlling reproductive hormone release and reproductive behaviours. Moreover, the neurohypophysial release of OT into the circulation is most efficiently provoked by various kinds of stimulations of genitals and the breast in different mammalian species.

Unlike humans, the majority of animals breed only during certain times of the year, and the expression of sexual behaviour is under strict endocrine regulation. Male and female animals mate only when circulating steroid hormone levels are at appropriate concentrations, and to influence behaviour, the steroid hormones must profoundly affect the neurotransmission in the brain. In contrast, humans and other primates display a phenomenon called "concealed ovulation" that may have played a role in the evolution of social structures.

Female Sexual Behaviour in Animals 1

In the female rat, the sexual behaviour is generally divided into proceptive (soliciting) and receptive (lordosis) responses, both of which are under the control of gonadal steroids. Many studies have shown that intracerebral infusion of OT into the hypothalamus facilitates both aspects of the sexual behaviour in estrogen-primed female rats. In many but not in all studies, progesterone administration was necessary for the OT-induced facilitation of lordosis behaviour. In particular, prolonged (3 days) priming with estradiol induced lordosis, even in the absence of progesterone. Furthermore, the frequency and/or duration of lordosis was differentially affected, dependent on the site of OT administration. It was therefore suggested that OT may control different aspects of lordosis. Surprisingly, low concentrations of OT could even suppress lordosis when infused into the lateral ventricle of female rats.

On the other hand, several experiments with OT antagonists confirmed that endogenous OT is essential for the expression of sexual behaviour. So, infusion of an OT antagonist into the medial preoptic area before progesterone inhibited sexual receptivity and increased rejection in female rats. Additionally, in nonovariectomised female rats in spontaneous behavioural estrus, intracerebroventricular injection of OT increased lordosis quotient and lordosis duration, whereas an OT antagonist prevented this OT-induced effect. Witt and Insel observed that significant OT antagonist effects were absent in females primed with estradiol alone, suggesting that the OT antagonist attenuates progesterone facilitation of female sexual behaviour. Surprisingly, these behavioural effects of OT antagonist administration did not appear immediately, but emerged only when antagonist was given with progesterone 4–6 h before behavioural testing. Two other observations support further evidence for the involvement of endogenous OT in the expression of female sexual behaviour. First, the immunoreactivity of the immediate early gene product Fos was induced in OT neurons after sexual activity in female rats. Second, intrahypothalamic infusions of antisense oligonucleotides directed against the OT receptor mRNA reduced lordosis behaviour as well as OTR in the VMN of estrogen-primed rats.

How could OT act to induce the expression of lordosis? Vincent and Etgen showed that steroid priming promotes an OT-induced norepinephrine release in the ventromedial hypothalamus of female rats. This occurs most probably by a peripheral mechanism, e.g., vaginocervical contraction, leading to increased excitability of ventromedial hypothalamus neuronal activity and the expression of lordosis. With respect to OT effects on sexual behaviour, only few studies were performed in species other than rats. In ovariectomised hamsters (*Mesocricetus auratus*) primed with estradiol, microinjection of OT into the medial preoptic area-anterior hypothalamus or the ventromedial hypothalamus induced sexual receptivity, whereas injection of a selective OT antagonist reduced the levels of sexual receptivity. OT injection into the hypothalamus also increased ultrasonic vocalisations that female hamsters use to alert and attract potential mates.

Male Sexual Behaviour 2

Male sexual behaviours and performances have also been mostly studied in rats and are measured, e.g., by changes in the frequency or durations of mounts, penile erections, intromissions, ejaculations, and yawnings. Yawning is considered to be a phylogenetically old, stereotyped event that occurs alone or associated with stretching and/or penile erection under different conditions.

OT was found to be one of the most potent agents to induce penile erection in rat, rabbits, and monkeys. Melis et al. reported that low doses of OT (e"3 ng) can elicit penile erection and yawning in rats following site-specific injection into the PVN, i.e., the nucleus which contains cell bodies of the majority of oxytocinergic neurons projecting to extrahypothalamic brain areas. Both OT effects were prevented by intracerebroventricular injection of OT antagonists as well as by electric lesions of the PVN. According to Argiolas and Melis, the oxytocinergic neurons facilitate yawning by releasing OT at sites distant from the PVN, i.e., the hippocampus, the pons, and/or the medulla oblongata when activated by the D1/D2 dopamine agonist apomorphine, *N*-methyl-D-aspartate, or OT itself. This activation was antagonised by opioid peptides that prevent the yawning response. Central administration of OT also significantly reduced the latency to achieve ejaculation and shortened the postejaculatory interval. This effect was more pronounced in aged male rats and might be related to decreases in circulating testosterone. Apomorphine- and OT- induced penile erection and yawning were endocrine dependent and are differentially modulated by sexual steroids.

It was hypothesized that OT induces sexual behavioural responses by increasing NO synthase activity in the cell bodies of oxytocinergic neurons projecting to extrahypothalamic brain areas, e.g., via intracellular [Ca^{2+}] increase and stimulation of the Ca^{2+}/calmodulin-dependent NO synthase. In agreement with this suggested mechanism, OT-induced penile erections were prevented by PVN injections of É-conotoxin, a blocker of N-type voltage-dependent calcium channels. The PVN is one of the richest brain areas containing NO synthase, and NO is a well-known mediator of penile erection, both peripherally and centrally. Accordingly, injections of NO donors into the PVN induced penile erections, and this response was prevented by the intracerebroventricular administration of an OT antagonist. Conclusively, OT may induce male sexual behaviours by activating NO synthase in the PVN. The concomitant production of NO could cause the activation of oxytocinergic neurons projecting to extrahypothalamic brain areas such as the hippocampus and/or spinal cord and control the male sexual function.

However, not all studies are in favour of a central facilitatory effect of OT on male sexual activity. OT produced an immediate cessation in sexual activity of male prairie voles and remained so for at least 24 h. In another study, centrally injected OT prolonged the postejaculatory refractory periods. These and some other findings are consistent with the hypothesis that OT could also play a role in sexual satiety. OT-immunoreactive neurons in the parvocellular subnuclei of the PVN project to a sexually dimorphic motor nucleus in the lower spinal cord which innervates the striated bulbocavernosus muscle. This muscle is responsible for penile reflexes, and it was suggested that this pathway is involved in the integration of penile reflexes with other aspects of male copulatory behaviour that are under hypothalamic control.

Sexual Behaviour in Humans 3

It is well documented that levels of circulating OT increase during sexual stimulation and arousal and peak during orgasm in both men and women. Murphy et al. measured plasma OT and AVP concentrations in men during sexual arousal and ejaculation and found that plasma AVP but not OT significantly increased during arousal. However, at ejaculation, mean plasma OT rose about fivefold and fell back to basal concentrations within 30 min., while AVP had already returned to basal levels at the time of ejaculation and remained stable thereafter. Men who took the opioid antagonist naloxone before self-stimulation had reductions in both OT secretion and the degree of arousal and orgasm. A recent study confirmed that also in women peak levels of serum OT were measured at or shortly after orgasm. Carmichael et al. reported that the intensity of muscular contractions during orgasm in both men and women were highly correlated with OT plasma level. This suggests that some of OT's effects may be related to its ability to stimulate the contraction of smooth muscles in the genital-pelvic area. Enhanced sexual arousal and orgasm intensity was reported in a woman during intranasal administration of OT. This response could be elicited only while she was taking daily doses of an oral contraceptive with estrogenic and progestogenic actions and might be caused through direct effects on sexual organs or sensory nerve sensitivity. Overall, beyond its peripheral effects on reproductive organs, OT might affect or sensitise cerebral neurons responsible for the cognitive feelings of orgasm and could thus serve as a physiological substrate for both sexual behaviour and performances in humans as well as in animals.

Maternal Behaviour B

The development of maternal care has been well studied in rats. Nulliparous females display little interest in infants and when presented with foster young will either avoid or cannibalise them. At parturition, however, a dramatic shift in motivation occurs and maternal behaviours such as nest building and retrieval of pups became established. Pedersen and Prange first demonstrated that injection of OT into the lateral ventricles of nulliparous ovariectomised rats induces maternal behaviour. The rapid onset of maternal behaviour in response to OT has been confirmed in several studies. However, it is important to note that OT is effective only to initiate the maternal behaviour, but not for the performance of maternal behaviour per se. So, when the females become maternal or enter into estrus, an OT antagonist had no effect. Moreover, steroid priming was found to be essential for the initiation of maternal behaviours in all cases so far studied. Subcutaneous injection of the NO donor sodium nitroprusside was shown to prolong parturition and to inhibit maternal behaviour in rats. Because OT was able to restore intrapartum maternal behaviour, NO was suggested to interfere with the initiation of maternal behaviour via blocking or diminishing the release of OT. Genital stimulations that are known to activate the oxytocinergic system could also provoke maternal behaviours as shown in ewes and rats. Genital stimulations may also influence olfactory function through the activation of afferent noradrenergic pathways in the olfactory bulbs. The olfactory bulb is in fact regarded as a critical site where OT may induce onset of maternal behaviour. OT originating in the hypothalamic PVN possibly acts there to decrease olfactory processing. This is in line with earlier findings according to which anosmia favours the establishment of maternal behaviour. However, the female gets a variety of sensory stimulations from contact with the pups so that the neural circuits responsible for postpartum maternal behaviour are rather complex. In addition, recent studies demonstrate that the paternal genome has aquired the ability to regulate maternal behaviour by a epigenetic mechanism called genomic imprinting, which results in the parent-of-origin-specific expression of only one allele of a gene. A striking impairment of maternal behaviour was observed in mice in which neurally expressed imprinted genes of paternal origin were mutated or deficient. One hypothesis to explain this phenomenon is that the paternal interest is best served by prolonged care and feeding of his progeny through maternal lactation.

The role of OT for maternal behaviour became questionable by studies with transgenic mice, since female mice genetically deficient in OT displayed an apparently normal maternal behaviour. This clearly demonstrates that OT itself is not essential for maternal behaviours in mice. Of course, this does not rule out that other OT-like ligands are present to activate the OT receptors or that in other species than mice, OT is essential to develop normal maternal behaviours.

In humans, OT-related maternal behaviours have not been the subject of any systematic studies so far. In earlier reports it was shown that breast-feeding within 1 h of birth, when OT levels are very high, supports a long-lasting mother-infant bond and has a beneficial effect on the development of the child.

Social Behaviour C

Parental care, nursing, social interaction, pair bonding, and mutual defence are prototypical mammalian species behaviours and have been important for the successfull evolution of mammals. Love and social attachments function to facilitate reproduction, provide a sense of safety, and reduce anxiety or stress. These social behaviours are clearly opposed to the ancient self-preservative behaviours. When these balanced social interactions are disturbed, e.g., by a stressor, the self-preservative, fight-flight response pattern takes priority. Recent studies in rodents suggest that the neurohypophysial hormones in concert with steroids are key components in the central mediation of complex social behaviours, including affiliation, parental care, sexual behaviour, mate guarding, and territorial aggression. OT has been reported to mediate aggressive and affiliative behaviours in several species. Moreover, disruption of the OT gene in mice showed reduced aggression behaviours in some tests.

Social behaviours have been intensively studied in North American species within the genus *Microtus* (voles), which exhibit diverse forms of social organisation ranging from minimally parental and promiscous (e.g., montane vole, *M. montanus*) to biparental and monogamous (e.g., prairie vole, *M. ochrogaster*) social structures. The highly social prairie voles usually form long-term monogamous relationships as a consequence of mating and thereby develop a variety of affiliative behaviours, such as parental care, increased physical contacts with each other, and the attacking of adult intruders as a putative mate guarding response. Because vaginocervical stimulation concomitant with the intense mating of the prairie voles activates the oxytocinergic system, OT was suggested as a good candidate for the facilitation of pair bonding in prairie voles. Moreover, in females that did not mate, intracerebroventricular but not peripheral administration of OT facilitated the formation of a pair bond. Conversely, intracerebroventricular injection of an OT antagonist before mating prevented partner preference formation in female prairie voles. This suggests that OT, released in response to mating behaviour, is sufficient and necessary for the female to form a pair bond to her mate. In males, however, AVP, but not OT, was shown to be critically involved in partner preference and selective aggression. In contrast, neither peptide appears to induce pair bonding in the promiscuos montane voles. Whereas OT had little effect on the social behaviour of montane voles, the same dose of AVP that induced aggression in the prairie vole increased self-grooming in the montane vole. Since the *Microtus*species share similar OT and AVP immunoreactive patterns but respond differently to exogenous peptide, the neuroendocrine differences between monogamous and polygamous species probably reside on the corresponding receptor molecules. In addition, the neuroanatomical distribution of OT and AVP receptors in several *Microtus* species suggested that the pattern of OT and AVP receptor binding is associated with social organisation. High OT receptor densities were found in the prelimbic cortex and nucleus accumbens of prairie voles, but not montane voles. Because these brain areas are involved in the mesolimbic dopamine reward pathway, it was hypothesized that in the female prairie voles OT released in response to mating activates this reward pathway and may condition the female to the odour of her mate. Furthermore, in the montane vole, which shows little affiliative behaviour except during the postpartum period, brain OT receptor distribution changed within 24 h of parturition, concurrent with the onset of maternal behaviour. This suggests that variable expression of the OT receptor in brain could be an important mechanism in evolution of species-typical differences in social bonding and affiliative behaviour.

The complexity of the neuropeptide's actions on social behaviour is illustrated by studies in the squirrel monkey, in which the effects of exogeneous OT depended on the prior social status of individuals. Dominant males responded to OT administration with increased sexual and aggressive behaviour, whereas subordinate males displayed more associative behaviours. The differential behaviours might be due to the higher serum testosterone levels in dominant animals.

Transgenic mice created with the 52 -flanking region of the prairie vole OT receptor gene demonstrated that sequencing in this region influences the pattern of expression within the brain. Promoter sequences of the prairie vole OT receptor and V1a receptor genes and the resulting species-specific pattern of regional expression provide a potential molecular mechanism for the evolution of pair-bonding behaviours and a cellular basis for monogamy. Recently, Young et al. reported that an AVP-induced increase in affiliative behaviours could be observed in mice that were transgenic for the prairie vole V1a receptor gene.

Interactions among OT, AVP, and glucocorticoids could provide substrates for dynamic changes in social behaviours, including those required in the development and expression of monogamy. Results from research with voles suggest that the behaviours characteristics of monogamy may be modified by hormones during development and may be regulated by different mechanisms in males and females. Nonnoxious sensory stimulation associated with friendly social interaction induces a response pattern involving sedation, relaxation, and decreased sympathoadrenal activity. It is suggested that OT released from parvocellular neurons in the PVN in response to nonnoxious stimulation integrates this response pattern at the hypothalamic level. The health-promoting aspect of friendly and supportive relationships might be a consequence of repetitive exposure to nonnoxious sensory stimulation. On the other hand, OT and AVP could also be involved in the pathophysiology of clinical disorders, such as autism, which is characterised by an inability to form normal social attachments.

Stress-Related Behaviour D

In both male and female rats, OT exerts potent antistress effects such as decreases in blood pressure, corticosterone/cortisone level, and increases in insulin and CCK level. After repeated OT treatment, weight gain could be promoted, and the healing rate of wounds increased. Furthermore, acute exposure of rats to immobilisation stress resulted in an increase in OT mRNA level. OT secretion was also increased by forced swimming in virgin and early pregnant rats. In addition, microdialysis experiments on male rats showed specific increases in both central and plasma OT in response to forced swimming and shaker stress. Thus the stimulated release of OT could function to facilitate the activation of the hypothalamic-pituitary-adrenal axis and increase glucocorticoid release. In contrast to most other OT-induced behavioural and physiological effects, the antistress effects could not be blocked by OT antagonists, suggesting that yet unidentified OT receptors may exist.

It is also well documented that stress-induced central release of OT can ameliorate the stress-associated symptoms such as anxiety. So, OT has anxiolytic properties in estrogen-treated females in mice and in rats. Estrogen-induced increases in OT binding density in the lateral septum may contribute to the facilitation of social interactions. Like many other anxiolytic compounds, OT acts as an antidepressant as shown in two animal models of depression. Moreover, the antinociceptive action of OT is consistent with its antidepressant action. Some of these effects may be mediated by an influence of OT on dopaminergic neurotransmission in limbic brain regions.

Because stress and anxiety can impair maternal caretaking and reduce milk ejection, reduced stress responsiveness during lactation is adaptive for both mother and infant. Accordingly, lactating women had reduced hormonal responses to exercise stress when compared with postpartum women who bottle-feed their infants. In addition, women with panic disorder can experience a relief of symptoms during lactation. Because OT levels increase in blood and CSF after nonnoxious sensory stimulation such as touch, light pressure, and warm temperature in both female and male rats, OT could not only be responsible for the antistress effects occurring during lactation but may also be involved in health-promoting effects of relationships, social contacts, and networks.

Some of the OT-induced effects may be explainable by the anxiolytic actions of OT. For example, OT acting as an anxiolytic may reduce the inhibition inherent in social encounters. Also, behavioural tests in the laboratory frequently involve the exposure of the animal to a novel environment, combined with exposure to an unfamiliar conspecific. These stimuli are likely to induce a stress response, and perhaps this anxiety is reduced by OT. So, a unifying principal in OT action in the brain may be to facilitate social encounters by reducing the associated anxiety. However, more studies are necessary to clearly distinguish between physiological and pharmacological effects.

Feeding and Grooming E

OT acts as a "satiety hormone" in animals since both peripherally and centrally administered OT reduces feeding. In addition, food and anorexia-inducing agents, such as CCK, lead to pituitary OT secretion and subsequently to reduced food intake. This suggests that both nausea and satiety activate a common hypothalamic oxytocinergic pathway that controls the inhibition of digestion. Arletti et al. reported that OT, whether given intraperitoneally or intracerebroventricularly, reduced food consumption and the time spent eating, and it increased the latency to the first meal in fasted rats. Pretreatment with an OT antagonist completely prevented the feeding inhibitory effect of OT, and per se increased food intake. In particular, hyperosmolality is a very potent stimulus for OT release, and central OT was observed to mediate osmolality-related inhibition of salt appetite but was not essential for Na^+-sensing mechanism in salt appetite in rats. While Arletti et al. reported that OT directly inhibited both food and water intake in rats, Olsen et al. observed a relatively small effect of OT on water intake when given in doses that significantly inhibited food intake. How could OT mediate its influence on the ingestive behaviours? PVN neurons in general, and OT projections in particular, could either act to modulate the activity of intrinsic brain stem reflex arcs or exert a direct control over vagal efferents that project to the gut and inhibit the gastric motility.

In mice, central administration of OT elicits dramatic behavioural excitation such as stress-induced escape, scratching, and grooming, particularly a pronounced self-grooming. Slow infusion of OT into the PVN also initiates self-grooming in rats. OT activates several aspects of grooming, preferentially genital grooming. The ability of OT to facilitate grooming involves activation of dopamine D1 receptor-mediated neurotransmission in the mesolimbic pathway as shown with knock-out mice lacking the D1 dopamine receptor. However, neurotransmitters other than dopamine (e.g., endorphins) may also play a supplementary role in neuropeptide-enhanced grooming in mice. Grooming is a behavioural response to stressors, and the induction of grooming might be part of a homeostatic mechanism for reducing anxiety and facilitation of pair bonds, whether between mother and infant or between mating partners, by increased transfer of odours between individuals.

Memory and Learning F

OT and AVP are considered to play a pivotal role in various aspects of learned behaviour. Overall, the concept emerged that AVP reinforces memory, whereas OT has just the opposite effect, namely, the attenuation of learning processes and memory. In particular, OT was shown to facilitate the extinction of avoidance reaction and to attenuate the storage of verbal memory. In cultured neurons, OT at concentrations over 1 M reduces the activity of NMDA receptors, thus impairing one of the major substrates for the induction of learning and memory. Regarding its impairing effects on some memory-related tasks, OT could possibly be involved in the forgetting of delivery pain in mothers. On the other hand, studies on rodents indicated that socially relevant behaviours are controlled in a more complex way by both AVP and OT released intracerebrally and argued against a simplistic view with respect to the memory-attenuating effects of OT. In this context, reports on the social memory in rats provide a good example. Social recognition can be measured in rats as the specific decrease in social investigation of a juvenile conspecies during the second encounter of the same individual. This social memory is based on the olfactory characteristics of the conspecies and is short lasting (<40 min.). OT had an impairing effect on the social memory in male rats. In another study, intraperitoneally injected OT to male rats significantly improved social memory. Depending on the dose, social memory was shown to be attenuated or facilitated by OT and derivatives as reported by Popik et al.. Interestingly, the active moieties for attenuation and facilitation of social memory reside in different parts of the OT molecule, e.g., the 5–7 and 8–9 region, respectively. However, the wide variety of observed effects has also led to the suggestion that OT has a more general effect on the cortical arousal rather than a specific effect limited to a certain stage of information processing.

Tolerance and Dependence to Opioids G

Drug tolerance, dependence, and addiction may involve neuroadaptive mechanisms related to learning and memory at cellular and systems levels. Chronic morphine administration alters the brain OT system, which suggests that OT might contribute to the behavioural, emotional, and neuroendocrine responses to opioids. Moreover, OT has been shown to inhibit the development of tolerance to morphine and to attenuate various symptoms of morphine withdrawal in mice. OT also decreased cocaine-induced locomotor hyperactivity and stereotyped behaviour in rodents. Dopaminergic neurotransmission and central OT receptors located in limbic and basal forebrain structures are probably responsible for mediating these various effects of OT in the opiate- and cocaine-addicted organism.

The influence of opioids on the secretory activity of OT neurons in the SON was studied in rats in more detail. The brain stem noradrenergic system that is activated, e.g., by peripheral administration of CCK, represents a major excitatory input to OT neurons. Morphine inhibits the OT neurons presynaptically via inhibition of the afferent noradrenergic input. When the opiate antagonist naloxone is administered locally in the SON, the inhibitory effect of morphine on CCK-stimulated norepinephrine release is reversed, and the firing rate of OT cells markedly increases. However, noradrenergic systems are not essential for the expression of morphine withdrawal excitation, since chronic neurotoxic destruction of the noradrenergic inputs to the hypothalamus did not affect the magnitude of withdrawal-induced OT secretion. Opioid receptors on OT neurons and/or on their afferent input may be activated by the endogeneous opioids pro-dynorphin and pro-enkephalin and are involved to restrain the secretory activity of OT cells. Interestingly, the efficiency of the endogeneous opioid restraint changes during pregnancy in rats. Therefore, Russell et al. have speculated that central opioid mechanisms could control OT neurons during parturition and can interrupt established parturition by inhibiting OT neuron-firing rate in disadvantageous environmental circumstances.

Central Disorders in Humans H

Obsessive-compulsive Disorder 1: Obsessive-compulsive disorder (OCD) includes cognitive and behavioural symptoms that are related to OT-associated behaviours. A possible role for OT in the neurobiology of a subtype of the OCD was suggested by the elevated CSF levels of OT and by the correlation between CSF OT levels and the severity of the disorder. Excessive grooming is often a component of OCD, and central OT administration induces high levels of self-grooming in animals. Periods of increased gonadal steroid secretion such as puberty and pregnancy are associated with a heightened risk for onset of OCD. Perhaps, gonadal steroids activate OT receptor during these periods. However, the results of another study do not support the hypothesis that OT might be a potential anticompulsive agent.

Eating Disorders: Anorexia and Prader-Willi Syndrome 2

The OT neurons of the PVN are good candidates for playing a physiological role in ingestive behaviour as "satiety neurons" in the human hypothalamus. In humans, OT appears not to be released in response to a meal. In women with anorexia nervosa, CSF levels of OT are reduced during the starvation phase of the illness and return to normal as the patients increase their food intake. Both hypoglycemia- and estrogen-induced OT increases were lower in underweight anorectic patients compared with normal controls. Anorectic subjects regained normal OT responsiveness to both stimuli after complete weight recovery. Reduced brain and plasma levels of OT during the starvation phase of the illness might contribute to the symptom profile of anxiety, loss of libido, amenorrhea, and increased activation of the hypothalamic-pituitary-adrenal hormonal stress axis.

In Prader-Willi syndrome, a genetic disorder characterised by gross obesity, insatiable hunger, hypotonia, hypogonadism, mental retardation, and a high risk for OCD, a 42% decrease in the number of hypothalamic OT neurons was observed. Moreover, there was a reduction in total cell number in the PVN, and the volume of the PVN-containing OT-expressing neurons was decreased by 54%. The lower number of OT neurons in the hypothalamic PVN is presumed to be the basis of the insatiable hunger and obesity of patients with the syndrome. Further evidence for hypothalamic and oxytocinergic dysfunction in Prader-Willi syndrome was provided by the finding that CSF OT was nearly twofold higher in patients with Prader-Willi syndrome compared with control subjects.

Depression and Schizophrenia 3

In postmortem samples from patients with depression, a 23% increase in the number of OT-immunoreactive neurons in the PVN of the hypothalamus was found. The putative higher rate of oxytocinergic activity in these patients is possibly associated with activation of the hypothalamic-pituitary-adrenal axis in these patients, since OT is known to potentiate the effects of CRH.

Increased concentrations of OT were observed in the CSF of schizophrenic patients, and they were higher in patients receiving neuroleptic treatment. However, Glovinsky et al. found no changes of the OT concentration in the CSF of schizophrenic patients either with or without neuroleptic medication. In another study, the basal level of OT-neurophysin was found to be threefold higher in the plasma of schizophrenics. So, impairments in the function of the OT system could play a role in the social deficits of schizophrenic patients, and OT is a candidate novel endogenous antipsychotic.

Neurodegenerative Diseases 4

OT injected into the hippocampus is reported to interfere with the formation of memory in experimental animals. In humans, high density of OT binding sites was observed in the nucleus basalis of Meynert, which is the major nucleus for cortical cholinergic neurons and degenerates in Alzheimer's disease. It has been reported that OT binding site density in the basal nuclei of Meynert decreases in the brains of patients suffering from Alzheimer-type senile dementia. OT concentration was found to be increased 33% in the hippocampus and temporal cortex of Alzheimer brains but was normal in all other regions examined. In two other sets of patients with Alzheimer dementia, the number of hypothalamic OT-expressing cells remained unaltered with aging. A slightly decreased number of OT-immunoreactive neurons was reported in the PVN of the hypothalamus in Parkinson's disease. However, more studies are needed to evaluate the role of OT for neurodegenerative disorders.

Concluding Remarks

In the past decade, the OT receptor structure has been elucidated, and considerable advances have been made in understanding the structure and function of the OT receptor system that has now been detected in many different tissues. Is there a further OT receptor subtype as suggested by some findings? It is certainly too early to exclude this possibility. On the other hand, many of the unexplained observations in the OT receptor field may result from complex cross-talks to poorly defined signalling cascades, interactions with receptor modulators like Mg^{2+} and cholesterol, and/or regulation by steroids via genomic and nongenomic pathways. Particularly, the functional dependence on steroids, a characteristic feature of the OT receptor system, is among the least understood. This is not surprising in view of the multiple targets of steroids and the novel facets of steroid receptor actions. For example, recent observations suggest that several steroid receptors can be activated by various agents in the absence of cognate hormone. To clarify the underlying mechanisms for both genomic and nongenomic steroid actions will therefore be of fundamental importance to understand the physiological regulation of the OT receptor system.

To some extent, most of the different actions of OT can be integrated in a concept according to which OT supports and facilitates the reproduction at several levels. In social living species like humans, affiliative behaviours are an essential component of reproduction. It will be therefore interesting to see whether in humans, mislocations of the OT receptor, naturally occurring mutations, or polymorphisms in the OT receptor gene can be correlated with (gender-specific?) physiological or behavioural deficits. In view of the widespread OT-related actions, OT antagonists may not only be regarded as promising candidates to prevent preterm labour and dysmenorrhea, but may together with OT agonists also prove useful for treatment of psychiatric illnesses such as anxiety, drug abuse, sexual dysfunctions, eating disorders, or autism.

Chapter 6

The Multifunctional Fish Gill

Fishes are aquatic vertebrates that are members of the largest and most diverse vertebrate taxon (~25,000 species), that dates back over 500 million years. They have evolved into three major lineages: Agnatha (hagfish and lampreys), Chondrichthyes (sharks, skates, and rays; usually referred to as elasmobranchs), and Actinopterygii (bony fishes, with teleosts being the most prevalent). Regardless of lineage, the majority of fish species uses the gill as the primary site of aquatic respiration. Aerial-breathing species may use the gill, swim bladder, or other accessory breathing organs (including the skin). The fish gill evolved into the first vertebrate gas exchange organ and is essentially composed of a highly complex vasculature, surrounded by a high surface area epithelium that provides a thin barrier between a fish's blood and aquatic environment.

The entire cardiac output perfuses the branchial vasculature before entering the dorsal aorta and the systemic circulation. The characteristics of the gill that make it an exceptional gas exchanger are not without trade-offs. For example, the high surface area of the gills that enhances gas exchange between the blood and environment can exacerbate water and ion fluxes that may occur due to gradients between the fish's extracellular fluids and the aquatic environment. In the past 50 years, it has become clear that the branchial epithelium is the primary site of transport processes that counter the effects of osmotic and ionic gradients, as well as the principal site of body fluid pH regulation and nitrogenous waste excretion. Thus the branchial epithelium in fishes is a multipurpose organ that plays a central role in a suite of physiological responses to environmental and internal changes. Despite the fact that fishes do have kidneys, the gill actually performs most of the functions that are controlled by pulmonary and renal processes in mammals. The purpose of this review is to integrate the latest morphological, biochemical, and molecular data (with appropriate references to historical studies) in an effort to delimit what is known and unknown about this interesting organ.

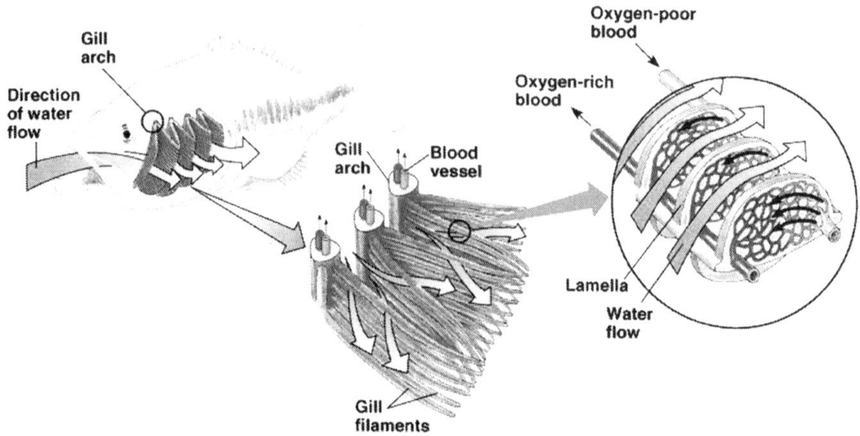

Figure: Schematic of the teleost fishgill.

Evolutionary Origin and External Structure

Gills of Protovertebrates and Modern Fishes

One of the defining characteristics of vertebrates is the presence of pharyngeal gill slits during some life stage (e.g., embryo, juvenile, or adult). In protovertebrates (i.e., urochordates, cephalochordates), the gills are relatively simple structures compared with those of extant fishes.

For example, in amphioxus (a cephalochordate), the gill slits are formed by numerous, vertically oriented gill bars in the pharynx. These gill bars contain an internal support rod (made of collagen), blood vessels, and neuronal processes; externally they possess a ciliated epithelium. Despite the presence of branchial blood vessels that connect the ventral aorta to the dorsal aorta, the gill bars of amphioxus do not have a large diffusing capacity for oxygen uptake and are not a likely site of respiratory gas exchange. Instead, the gill bars play a prominent role in feeding by filtering food particles from water that is passed through the gill slits by the ciliated gill bar epithelium.

In contrast to the specialised feeding device of protovertebrates, the gills of modern fishes have evolved into an anatomically complex, multifunctional tissue with discrete external (i.e., epithelial) and internal (i.e., circulatory and neural) elements. The gills of fishes are located near the head region and are composed of several paired gill arches on the pharynx. Anchored to the gill arches is a complex arrangement of epithelial, circulatory, and neural tissues, which will be described in the next two sections.

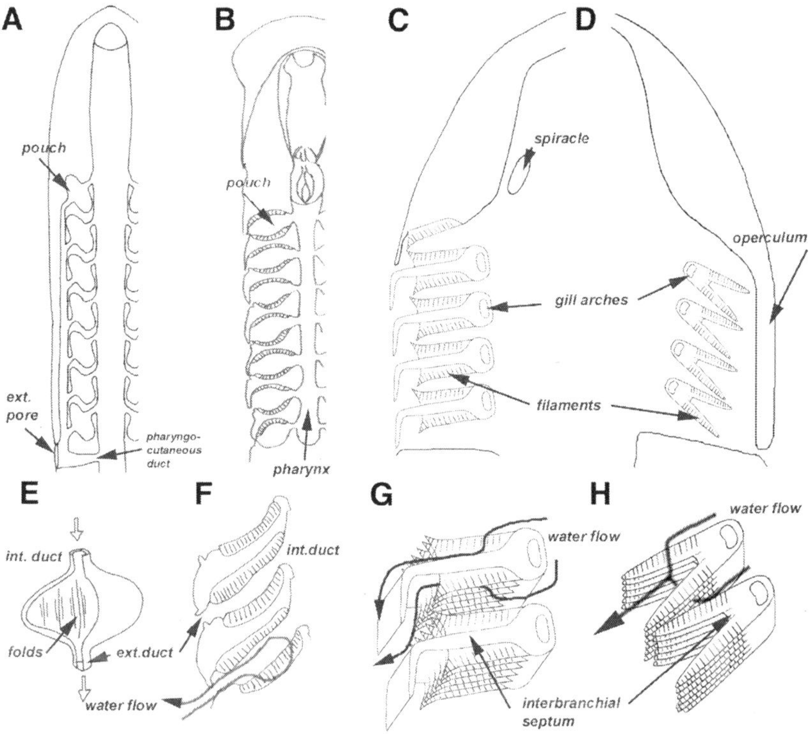

Figure: Generalised schematics of gills and associated pouches/arches in hagfishes (*A* and *E*), lampreys (*B* and *F*), elasmobranchs (*C* and *G*), and teleosts (*D* and *H*). All schematics are oriented anterior to posterior, with oral opening at top.

External Structure of Fish Gills

Gross Anatomy

The general anatomy of the gills varies among the three extant evolutionary lineages of fishes, but a simplified arrangement can be generalised for elasmobranchs and teleosts (agnathans will be described separately). Typically, cartilaginous or bony support rods (gill rays) radiate laterally from the medial (internal) base of each gill arch, and connective tissue between the gill rays forms an interbranchial septum. The interbranchial septum supports several rows of fleshy gill filaments (termed hemibranchs) that run parallel to the gill rays on both the cranial and caudal sides of a gill arch. Gill filaments are the basic functional unit of gill tissue, and their structure is described in section II*B2*.

A set of cranial and caudal hemibranchs from the same arch is referred to as a holobranch. In most elasmobranchs, four pairs of holobranchs are present in the branchial chamber, with only a pair of caudal hemibranchs on the first gill arch. The interbranchial septum of elasmobranch holobranchs extends from the base of each respective gill arch to the skin and forms distinct external gill slits. The typical path of water flow through the elasmobranch gills is for water to enter the pharynx via the mouth or spiracles (cranial valves for water entry), then pass over the gill filaments and follow the interbranchial septum until the water exits via the gill slits.

In teleosts, only four pairs of holobranchs are present, and the interbranchial septum is much reduced compared with elasmobranchs. The septum usually only extends to the base of the filaments, and thus the filaments of teleosts are much more freely moving than those of elasmobranchs. No distinct external gill slits are found in teleosts, but a thin, bony flap called the operculum externally protects the branchial chamber. In teleosts, water enters the pharynx from the mouth, then passes over the filaments and follows the inner wall of the operculum until it exits via a caudal opening of the operculum.

Among agnathans, the gills of adult, parasitic lampreys have a similar overall organisation to elasmobranch gills, except no gill rays are present in lampreys, and the skeletal base of the gill arch in lampreys is situated external (lateral) to the hemibranchs; in elasmobranchs (and teleosts), the base of the gill arch is medial to the hemibranchs.

In addition, the interbranchial septa of lampreys are slightly concave in shape, which results in the cranial and caudal hemibranchs of adjacent holobranchs forming pouchlike structures with discrete internal and external branchiopores or ducts that open to the pharynx and environment, respectively. Lampreys have six pairs of holobranchs, with only a caudal and cranial hemibranch on the first and last gill arch, respectively; this results in a total of seven paired gill pouches. Water flow through lamprey gills is rather diverse compared with other fishes, because, during feeding, lampreys attach to their prey by oral suction, and therefore the typical mouth to gill water path is not available.

Through contractions of the extensive musculature that surrounds the pouches, water is moved into and out of the external ducts in a tidal manner to irrigate the gill filaments, which allows branchial respiration to occur while feeding. The flow of water through lamprey gills can also be modified to pump water from the external ducts, through the gills, and into the pharynx via the internal ducts. This water path may be used to clean the pharynx, or it may assist in the removal of the lamprey from its prey.

The gills of hagfish (the other agnathan lineage) have a rather unique gill anatomy and organisation compared with those of other fishes. The gills are composed of 5–14 pairs of lens-shaped pouches that have discrete incurrent and excurrent ducts that connect to the pharynx and the environment (either directly or indirectly via a common pore, depending on species), respectively. Unlike other fishes, no well-developed skeletal structures (e.g., arches) are associated with the pouches.

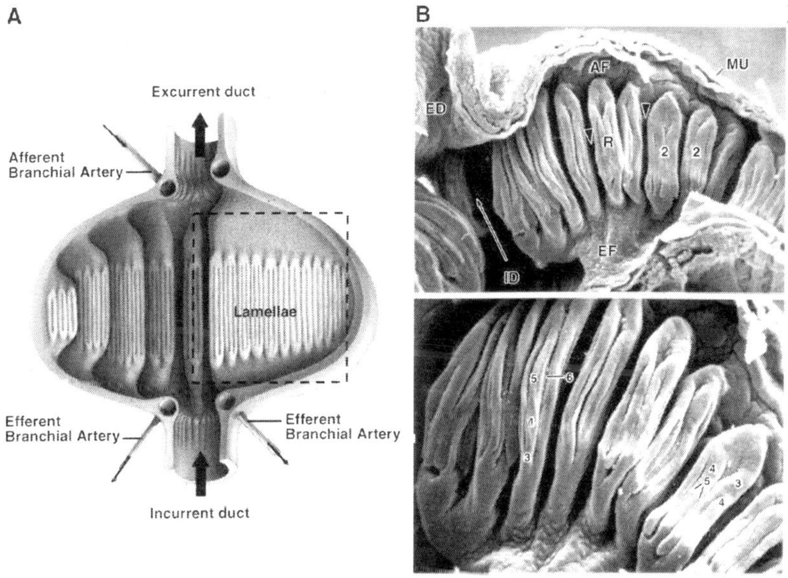

Figure: Anatomy of hagfish gills. *A*: schematic of a longitudinal cut through a gill pouch from the Atlantic hagfish, with a lateral perspective of a primary gill fold (filament) and its lamellae (boxed area). Note radial arrangement of additional filaments around the pouch. Large arrows indicate direction of water flow; small arrows indicate direction of blood flow. *B*: scanning electron micrographs of a gill filament from the Pacific hagfish, comparable to boxed area in *A*. Top micrograph (x30) shows an overview of a filament, with afferent (AF) and efferent (EF) regions, and respiratory lamellae (R) with second-order folds (2). MU indicates muscular layer around the pouch. Arrow indicates flow of water through pouch from incurrent duct (ID) to excurrent duct (ED); arrowheads indicate flow of blood across filament. Bottom panel (x70) reveals higher order folds of the lamellae, i.e., third- (3), fourth- (4), fifth- (5), and sixth-order (6) folds.

From the internal wall of each pouch, several extensive epithelial folds arise that span the lateral dimension of the gill pouch and extend towards the centre of the pouch. The folds are radially arranged around the mediolateral axis of the pouch. Despite their atypical arrangement and morphology, these folds are considered to be comparable to the gill filaments of other fishes.

The path of water through hagfish gills resembles that of other fishes, except that water enters the pharynx through a nasal opening instead of the mouth. Water then enters the pouches via their incurrent ducts, passes the gill filaments, and exits the pouches via excurrent ducts. In addition to this path, hagfish can irrigate the pharynx via a pharyngocutaneous duct on the left side of the animal that is caudal to the last gill pouch and connects the pharynx directly to the environment. This unique duct may be used to irrigate the pharynx and gill pouches while the animal is feeding, which usually involves burying the head region within a prey item.

Gill Filament Anatomy

In most fishes, gill filaments are long and narrow projections lateral to the gill arch that taper at their distal end. Each filament is supplied with blood from an afferent filamental artery (AFA) that extends along the filament. Blood in this vessel also travels across the filament's breadth through numerous folds on the dorsal and ventral surfaces of the filament-termed lamellae, which are perpendicular to the filament's long axis. Blood that crosses the lamellae drains into an efferent filamental artery (EFA) that runs along the length of the filament and carries blood in the opposite direction to that in the AFA. Blood flow through the gills and gill filaments is described in more detail in section IIIA.

Lamellae are evenly distributed along a filament's length, and the spaces between lamellae are channels through which water flows. A closer look at an individual lamella reveals that it is essentially composed of two epithelial sheets, held apart by a series of individual cells, termed pillar cells. The spaces around the pillar cells and between the two epithelial layers are perfused with blood, flowing as a sheet, not through vessels per. Lamellae dramatically increase the surface area of the gill filament epithelium and result in a small diffusion distance between the blood that perfuses each lamella and the respiratory water. Moreover, blood flow through the lamellae is countercurrent to water flow between them. Therefore, the lamellae are well-suited for gas exchange, but are also well-suited for diffusive losses or gains of ions and water to/from the environment.

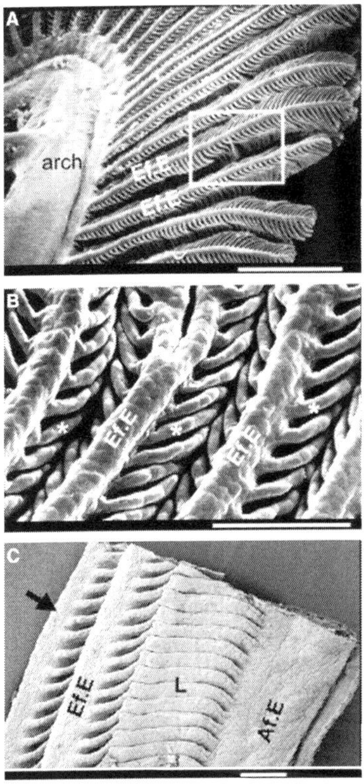

Figure: Scanning electron micrographs of gill filaments. *A*: filaments from a teleost (gulf toadfish, *Opsanus beta*), which radiate off a gill arch. Leaflike lamellae branch off the filaments (e.g., boxed area). Note the efferent edge (Ef.E) of the filaments, which is facing up; the afferent edge is in the depth of the figure. Bar = 100 μm. [From Evans (181).] *B*: high magnification view of filaments from an agnathan (pouched lamprey, *Geotria australis*), with the Ef.E of the filaments facing up. An asterisk indicates a lamella on each filament. Bar = 100 μm. [From Bartels et al. (29).] *C*: side profile of a filament from an elasmobranch (ocellated river stingray, *Potamotrygon motoro*) showing the afferent edge (Af.E), Ef.E, and lamellae (L). Arrow indicates the Ef.E of an underlying filament. Bar = 100 μm. (From P. M. Piermarini and D. H. Evans, unpublished micrograph.)

The description of the gill filament (and lamellae) is applicable to mostfish groups (i.e., elasmobranchs, teleosts, and lampreys), but not hagfish. As with their gill gross anatomy, the gill filament anatomy of hagfish is unique compared with other fishes. Each filament is supplied with blood from an afferent radial artery that begins near the excurrent duct of a pouch. Blood traverses medially across the filament where it encounters lamellae that are oriented in the same plane as the filamental blood flow; lamellae are oriented perpendicular to filamental blood flow in other fishes. Blood that flows through the lamellae is collected by an efferent radial artery near the incurrent duct of a pouch. Blood flow through hagfish gills and gill filaments is described in more detail in section IIIA.

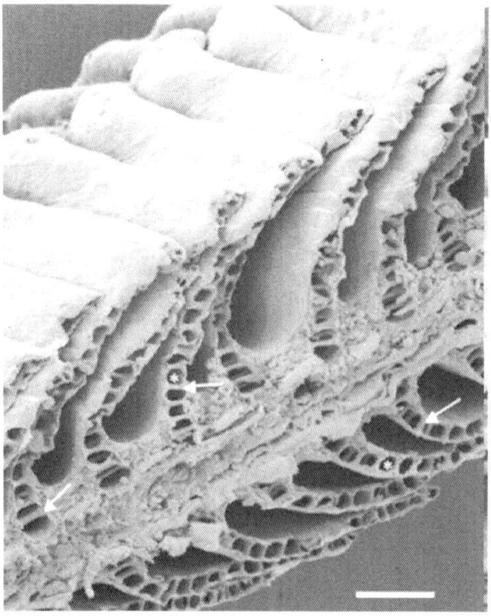

Figure: Scanning electron micrograph of a longitudinal cut through a gill filament from an elasmobranch (Atlantic stingray, *Dasyatis sabina*). Asterisks indicate lamellar blood spaces formed by pillar cells (arrows). Bar = 50 μm. (From Piermarini and Evans, unpublished micrograph.)

Lamellae of hagfish gills are much more complex than those of other fishes. For example, lamellae begin as secondary folds of the gill filament and continue to fold to form several higher order folds. These folds increase the surface area of the filament and, similar to lamellae in otherfishes, the folds are relatively thin structures composed of epithelial sheets separated by pillar cells. Regardless of the degree of folding, blood flow through the lamellae remains in the same direction (i.e., lateral to medial), which is countercurrent to water flow through the pouch. Thus, like other fishes, the lamellae of hagfish gills are well-suited for gas exchange.

In all fishes, the region of the filament that contains the afferent blood supply is commonly referred to as the afferent edge, whereas the region that collects efferent blood is referred to as the efferent edge. These two terms are synonymous with trailing edge and leading edge, respectively, relative to water flow across the filament. In hagfish, the additional term of *respiratory region* or*respiratory zone* is sometimes given to the lamellae and their higher order folds.

The Gill Epithelium

The epithelium that covers the gill filaments and lamellae provides a distinct boundary between a fish's external environment and extracellular fluids and also plays a critical role in the physiological function of the fish gill. The gill epithelium is composed of several distinct cell types, but primarily consists of pavement cells (PVCs) and mitochondrion-rich cells (MRCs), which comprise >90% and <10% of the epithelial surface area, respectively.

Pavement Cells

Although PVCs cover the vast majority of the gill filament surface area, they are largely considered to play a passive role in the gill physiology of most fishes. PVCs are assumed to be important for gas exchange because they are thin squamous, or cuboidal, cells with an extensive apical (mucosal) surface area and are usually the primary cell type that covers the sites of branchial gas exchange, i.e., the lamellae.

The apical membrane of PVCs is characterised by the presence of microvilli and/or microplicae (microridges), which often have elaborate arrangements that vary between species. These apical projections likely increase the functional surface area of the epithelium and may also play a role in anchoring mucous to the surface. Typically, PVCs do not contain many mitochondria but may be rich in cytoplasmic vesicles or have a distinct Golgi apparatus. In agnathans and elasmobranchs, PVCs possess subapical secretory granules or vesicles that contain mucous and fuse with the apical membrane. The intercellular junctions between PVCs and adjoining cells are extensive or multistranded, which makes the junctions "tight" and presumably relatively impermeable to ions.

In freshwater teleost gills, evidence suggests that some PVCs may play an active role in ion uptake and acid-base transport by the gills. For example, an ultrastructural study of the fresh water, brown bullhead catfish (*Ictalurus nebulosus*) gill epithelium demonstrated the presence of studded vesicles near the apical membrane of PVCs that resembled vacuolar-proton-ATPase (V-ATPase)-rich vesicles found in other ion regulatory epithelia (e.g., turtle urinary bladder and mammalian renal tubules). Subsequent immunohistochemical and biochemical studies in rainbow trout (*Onchorynchus mykiss*) and tilapia (*Oreochromis mossambicus*) have verified the presence of V-ATPase in PVCs. Moreover, recent studies have isolated a subpopulation of mitochondrion-rich PVCs (MR-PVCs) from rainbow trout gills that are V-ATPase-rich. These MR-PVCs appear to play an active role in sodium uptake and acid extrusion.

Mitochondrion-rich Cells

In contrast to PVCs, MRCs occupy a much smaller fraction of the branchial epithelial surface area, but they are considered to be the primary sites of active physiological processes in the gills. Whereas PVCs are found in all regions of gill filaments, MRCs are usually more common on the afferent (trailing) edge of filaments, as well as the regions that run between individual lamellae, termed the interlamellar region. Also, MRCs are usually not found on the epithelium covering the lamellae; however, certain environmental conditions are associated with the presence of lamellar MRCs in some species.

The ultrastructure and function of MRCs are highly variable among the three extant lineages of fishes, and therefore specific details on MRCs in agnathans, elasmobranchs, and teleosts will be described separately. However, a few general trends exist that are applicable to all groups. For example, MRCs are large ovoid-shaped cells, and as their name suggests, they have high densities of mitochondria in their cytoplasm, relative to PVCs. In addition, MRCs are highly polarised cells, with their apical and basolateral cell membranes having distinct morphologies and transport-protein expression profiles.

Teleosts: The ultrastructure of teleost MRCs has been studied extensively in several species, and numerous, extensive reviews exist on the subject. Teleost MRCs are often referred to as "chloride cells" in the literature, which is attributed to their NaCl secretory function in seawater teleosts. In freshwater teleosts, MRCs are also present, but they are characterised by different forms and functions than MRCs of seawater teleosts, and will be described separately.

One of the most striking characteristics of MRCs in seawater teleosts is the presence of an intricate tubular system that is formed by extensive invaginations of the basolateral membrane. This tubular/membranous labyrinth extends throughout most of the cytoplasm, where it closely associates with mitochondria and is the site of expression for the active transport enzyme Na^+-K^+-ATPase, which indirectly energizes NaCl secretion by these cells. The apical membrane of MRCs is concave and recessed below the surface of surrounding PVCs to form an apical pore or crypt that is shared with other MRCs. These crypts have been described as "potholes" in a "cobblestone street" of PVCs that expose the apical surface of MRCs to the ambient environment. The apical membrane itself is sparsely populated with short microvilli, and overall is morphologically unspecialised. However, the subapical cytoplasm contains a tubulovesicular system comprised of numerous vesicles and tubules that shuttle to and fuse with the apical membrane; these vesicles may be responsible for trafficking of the cystic fibrosis transmembrane regulator chloride channel to the apical membrane of MRCs, as seen by Marshall et al.. Intercellular junctions between MRCs and adjacent PVCs are extensive (multistranded) and are considered to be "tight," forming a relatively impermeable barrier to ions.

In seawater teleosts, MRCs exist in multicellular complexes with other MRCs and accessory cells (ACs). Similar to MRCs, ACs contain numerous mitochondria, but ACs are much smaller, with a less-developed tubular system and lower expression of Na^+-K^+-ATPase, relative to MRCs. The true nature of ACs is debatable; for example, in some species (e.g., brown trout; *Salmo trutta*) ACs appear to be a well-defined, discrete cell type, whereas in others (e.g., tilapia) evidence suggests that ACs are just a developmental stage of MRCs. Regardless, in seawater teleosts, the multicellular complexes form an apical crypt shared by the apical membranes of ACs and MRCs. Cytoplasmic processes of ACs extend into the apical cytoplasm of MRCs to form complex interdigitations. Importantly, these interdigitations form junctions between MRCs and accessory cells that are not extensive and are considered to be leaky to ions, thus providing a paracellular route for Na^+ extrusion.

It should be noted that high densities of MRCs that are ultrastructurally and functionally identical to MRCs of the seawater teleost gill epithelium are found in the opercular epithelium of killifish (*Fundulus heteroclitus*) and tilapia and the jawskin epithelium of the longjaw mudsucker. These flat epithelial sheets have been extremely valuable model systems for deciphering the mechanisms of ion transport in seawater teleost MRCs.

In freshwater teleosts, MRCs are primarily found on the afferent edge and interlamellar region of the gill filament, but lamellar MRCs are also present in some species. In general, the MRCs of freshwater teleosts share some ultrastructural characteristics with those from seawater teleosts, such as the extensive basolateral membrane infoldings that form a tubular system associated with mitochondria, and the subapical tubulovesicular system. However, important differences exist. For example, the ACs and multicellular complexes associated with MRCs in seawater teleosts are not common in freshwater teleosts; the MRCs often occur singly in freshwater teleost gill epithelium, intercalated between PVCs. Also, freshwater teleost MRCs form extensive, multistranded intercellular junctions with surrounding PVCs or other MRCs to form a relatively impermeable barrier to ions. The apical membrane of freshwater MRCs is usually flush with or noticeably protruded above adjacent PVCs and contains distinct patterns of microvilli or microridges. However, in some species (e.g., tilapia, mangrove killifish, *Rivulus marmoratus*) apical crypts still exist. Although the basolateral tubular system still occurs in MRCs of freshwater teleosts, this structure appears to be less developed compared with seawater MRCs.

In several species of freshwater teleosts, distinct morphological MRC subtypes have been detected in the gills (e.g., Atlantic salmon, *Salmo salar*; brown trout; guppies, *Poecilia reticulata*; loaches, *Cobitis taenia*; grudgeon, *Gobio gobio*; and Nile tilapia, *O. niloticus*). Pisam and colleagues have described two MRC subtypes (á and β) based on the degree of osmium staining (dark or light) in the cytoplasm, apical membrane morphology and associated subapical structures, extent of tubular system, cell shape, and anatomic location on the filament. The á-MRCs have light cytoplasmic staining, usually a smooth apical membrane associated with numerous subapical vesicles, a well-developed tubular system, an elongated shape, and are found at the base of lamellae where they may associate with an accessory cell (e.g., in Atlantic salmon and tilapia). In contrast, , β-MRCs have dark cytoplasmic staining, usually have complex apical membrane projections associated with an extensive subapical tubulovesicular network, a less-developed tubular system, an ovoid shape, and are found in the interlamellar region. In the brown trout,β-MRCs are associated with an accessory cell. Upon acclimation to seawater, the á-MRCs proliferate into the typical seawater teleost MRC, whereas the β-MRCs degenerate. Specific functional roles of á- and -MRCs in freshwater teleosts have yet to be identified.

Recent studies using immunohistochemistry and cell isolation techniques have corroborated and extended some of Pisam and colleagues' findings. For example, immunohistochemical studies of Na^+-K^+-ATPase in guppy and chum salmon (*Oncorhynchus keta*) gills corroborated the proliferation of á-MRCs upon seawater acclimation and also demonstrated that the á- and β-MRCs are rich and poor in Na^+-K^+-ATPase, respectively, which is consistent with the less-developed tubular system of β-MRCs. With the use of a Percoll density gradient and subsequent cell sorting to isolate MRCs from gill suspensions in Japanese eels (*Anguilla japonica*), two MRC populations have been identified in freshwater eels that correspond to the á- and -MRCs, and a decreased percentage of β-MRCs (with a corresponding increased percentage of á-MRCs) has been described upon seawater acclimatio.

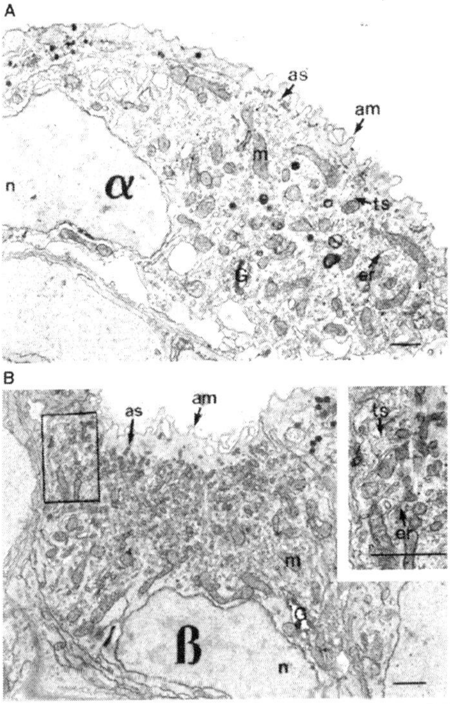

Figure: Transmission electron micrographs of α- and β-MRCs from a freshwater teleost (brown trout, *Salmo trutta*). *A*: α-MRC with few, small apical structures (as) that are primarily located below the apical membrane (am) and a well-developed tubular system (ts) that associates with parts of the endoplasmic reticulum (er). *B*: -MRC with many, large apical structures and a less-developed tubular system, compared with the α-MRC. In both panels, also note mitochondria (m), nuclei (n), and Golgi bodies (G). Bar = 1 μm.

This cell sorting technique has been used more recently to isolate, purify, and characterise two populations of MRCs (based on binding to peanut lectin agglutinin) in the gill epithelium of freshwater rainbow trout. The lectin-binding cells (PNA$^+$) have ultrastructural characteristics typical of teleost MRCs (e.g., a tubular system and tubulovesicular system), and the cells that did not bind lectin (PNA$^-$) are actually MR-PVCs. Reid et al. have referred to the PNA$^+$ MRCs and MR-PVCs as β- and á-MRCs, respectively, based on functional analogy to type B and type A intercalated cells of the mammalian nephron, respectively. However, it is not known if either of the above

MRCs in the rainbow trout directly correspond to the morphological β- and á-MRCs described by Pisam and colleagues in other teleosts.

A recent ultrastructural and immunohistochemical study on the gills of a euryhaline killifish (*F. heteroclitus*) has detailed the acute and chronic changes in MRCs associated with transfer from seawater to freshwater environments. In this study, it was found that MRCs from seawater killifish first morphologically and immunochemically convert to freshwater MRCs and then are gradually replaced by newly generated freshwater MRCs. Among the acute changes noted were *1*) a transformation of a recessed, concave apical membrane with few microvilli to a protrusive, convex apical membrane with extensive microvilli, within 7 days of freshwater exposure; *2*) degeneration of ACs and multicellular complexes within 12 h of exposure to fresh water; and *3*) loss of apical CFTR immunoreactivity within 24 h of exposure to freshwater.

Elasmobranchs: Despite a few recent studies in the spiny dog fish (*Squalus acanthias*), the ultrastructure of elasmobranch MRCs in general has been little studied. All ultrastructural descriptions of elasmobranch MRCs have been conducted on primarily marine species, and the ultrastructure of MRCs in the gills of freshwater elasmobranchs has not been examined to date. Elasmobranch MRCs are usually found on the afferent edge of the filament and between lamellae and exist singly in the epithelium. However, numerous MRCs have been identified on gill lamellae of an elasmobranch living in fresh water (Atlantic stingray, *Dasyatis sabina*). The tight junctions between elasmobranch MRCs and adjacent PVCs are multistranded, which suggests that the junctions are relatively impermeable to ions. In elasmobranch MRCs, the tortuous basolateral tubular system seen in marine teleost MRCs is lacking. However, the basolateral membrane of elasmobranch MRCs does have moderate infoldings, and they are likely the site of Na^+-K^+-ATPase expression in some MRCs and V-ATPase expression in other MRCs.

The subapical cytoplasm of elasmobranch MRCs is characterised by a tubulovesicular system that contains numerous vesicles, and the apical membrane is characterised by dense clusters of microvilli that can exist in different morphologies (Piermarini and Evans, unpublished data). This may suggest that different functional MRC subtypes are present in the elasmobranch gill epithelium, and recent data suggest that at least two immunohistochemically distinct populations of MRCs exist in the Atlantic stingray and spiny dogfish gills and that these cells are likely involved with distinct ion regulatory functions.

Agnathans: The ultrastructure of MRCs in agnathans varies considerably between hag fish and lampreys. In two hagfish species that have been examined (Atlantic hagfish, *Myxine glutinosa*; Pacific hagfish, *Eptatretus stouti*), MRCs are primarily found on the afferent edge of gill filaments. In contrast to marine teleost MRCs, hagfish MRCs exist singly in the epithelium and the tight junctions between MRCs and adjacent PVCs are extensive, which suggests they are impermeable to ions. Similar to marine teleost MRCs, hagfish MRCs have an extensive intracellular tubular system that is continuous with the basolateral membrane and is closely associated with mitochondria. This tubular system is likely the site of Na^{+}-K^{+}-ATPase expression, but high-resolution localisation data are still lacking. The apical membrane of hagfishMRCs contains microvilli, and the subapical region of the cytoplasm has a tubulovesicular system with densely packed vesicles that fuse with the apical membrane

In the gills of adult, anadromous lamprey species (e.g., pouched lamprey, *Geotria australis*; sea lamprey, *Petromyzon marinus*; river lamprey, *Lampetra fluviatilis*) ultrastructural studies have identified two distinct types of MRCs that are associated with freshwater and seawater life-stages. When in fresh water, lamprey gills possess a MRC subtype that is only found in freshwater individuals (intercalated MRC or FW-MRCs) and another MRC subtype that is found in both freshwater and seawater individuals (chloride cells or SW-MRCs). The FW-MRCs are found in the interlamellar region and the afferent edge of filaments where they occur singly in the epithelium, intercalated between PVCs and SW-MRCs. The basolateral membrane of FW-MRCs is not highly convoluted and is not associated with an elaborate tubular system. The FW-MRC apical membrane is characterised by numerous microridges that form distinct patterns, and the subapical cytoplasm is rich in vesicles that can fuse with the apical membrane. Additionally, freeze-fracture studies have found particles in the apical membrane of FW-MRCs that are associated with the presence of V-ATPase in other epithelia. Given these morphological properties and the occurrence of FW-MRCs only in freshwater lampreys, it has been suggested that this cell is involved with active ion uptake.

The SW-MRCs (or chloride cells) of lampreys are primarily found in the interlamellar region, but also occur on the afferent edge of the filament. They are laterally compressed in the longitudinal plane of the filament and line up next to one another for almost the entire length of the interlamellar region. Ultrastructurally, SW-MRCs share more similarities to marine teleost MRCs than to FW-MRCs of lampreys. For example, SW-MRCs have extensive basolateral membrane infoldings that form an intracellular tubular system, which closely associates with mitochondria. Additionally, SW-MRCs have a subapical tubulovesicular system composed of vesicles. In contrast to marine teleost MRCs, the apical membrane of lamprey SW-MRCs does not form an apical pit, but it does have relatively dense or sparse clusters of microvilli compared with surrounding cells, depending on salinity. Ultrastructural differences exist between the SW-MRCs of freshwater and seawater lampreys. For example, the tight junctions between SW-MRCs in freshwater lampreys are extensive, but upon acclimation to seawater the junctions become less extensive and are presumably leaky to ions.

The apical membrane morphology of SW-MRCs also varies with salinity. In freshwater lampreys the apical membrane surface area is restricted due to surrounding pavement cells that cover much of SW-MRC's apical surface, which results in a dense packing of apical microvilli on SW-MRCs. However, in seawater lampreys or freshwater lampreys recently acclimated to seawater, the PVCs appear to retract and expose more of the SW-MRCs apical surface, which becomes relatively flat with few microvilli. Finally, the tubular system of SW-MRCs appears to be more organised and robust in seawater lampreys compared with freshwater lampreys. All of these differences are consistent with a more prominent and active SW-MRC in the gills of seawater lampreys, which suggests they are involved with active NaCl secretion.

Bibliography

Anderson, A. Trevor: *Fish Nutrition in Aquaculture*, Springer, Delhi, 2009.

Bailey, MT: *A-Z of Tropical Fish Diseases and Health Problems*, Howell Book House, New York, 2001.

Balfour, F. M.: *A Monograph on the Development of Elasmobranch Fishes*: Macmillan, London, 1878.

Barlow, S.M.L: *Chemical and Biological Assay Procedures for Lysine in Fish Meals*, J.Sci.Food Agric., 1984.

Basant, Puri K.: *The Natural Way to Beat Depression: the groundbreaking discovery of EPA to change your Life,* Hodder and Stoughton, London, 2004.

Boserup, E.: *The Conditions of Agricultural Growth: The Economics of Agrarian Change under Population Pressure*, Aldine Press, Chicago, 1965.

Bowman, S.J.: *Digest of Statistics of IAFMM*, Potters Bar, U.K., IAFMM, 1984.

Bruno D.W., Poppe T.T., *A Color Atlas of Salmonid Diseases*, Academic Press, London, 1996.

Cech, JJ and Moyle, PB: *Fishes, An Introduction to Ichthyology*, Benjamin Cummings, NY, 2003.

Chow, Ching Kuang: *Fatty Acids in Foods and Their Health Implications*, Routledge Publishing, New York, 2001.

Clover, Charles: *The End of the Line: How Overfishing is Changing the World and What we Eat*, Ebury Press, London 2004.

Ferguson, H.W.: *Systemic Pathology of Fish*, Iowa State Press, Ames, Iowa, 1989.

Geran, James: *The Proper Care of Goldfish*, TFH Publications, NJ, 2000.

Hamm, Wolf: *Trans Fatty Acids*, Blackwell, NY, 2008.

Haskell, B.D.: *Ecosystem Health: New Goals for Environmental Management*, Island Press, Washington, D.C., 1992.

Hasler, A.D.: *Orientation and Fish Migration*, Academic Press, New York, 1971.

Kulkarni, G.K.: *Fisheries and Fish Toxicology*, APH, Delhi, 2006.

Lloyd, G.: *Mammals of the Pacific States,* Stanford University Press. 1965

Maddy Hargrove: *The Discus: An Owner's Guide to Happy Healthy Fish*, Howell Books, NY, 1999.

Mahadevan, R.K. and Selvamani, B.R.: *Fish and Fishery Culture: Assessment and Evaluation*, Campus Books International, Delhi, 2008

Malvee, Sangeeta: *Fish Genetics*, SBS Pub, Delhi, 2008.

Martin, A.M.: *Fisheries Processing: Biotechnological Applications*, Chapman and Hall, Delhi, 2009.

Mathur, A.K. and Sharma, L.L.: *Hand Book of Freshwater Ornamental Fishes,* Yash Pub house, Delhi, 2006.

Mehta, Varun: *Fisheries and Aquaculture Biotechnology*, Campus Books International, Delhi, 2006.

Mori, Fumitashi and Sakamoto, Yohei: *Aquarium Fish of the World: The Comprehensive Guide to 650 Species*, Chronicle Books, UK, 1991.

Moyle, PB and Cech, JJ: *Fishes, An Introduction to Ichthyology*, Benjamin Cummings, NY, 2003.

Nelson, J. S.: *Fishes of the World*, John Wiley & Sons, Inc.. , NY, 2006.

Petr, Tomi: *Fisheries in Irrigation Systems of Arid Asia*, Daya, Delhi, 2007.

Pollard, D. A. *Freshwater Fishes of South-eastern Australia*. McDowell, R. M. Reed, Sydney. 1980.

Ponniah, A.G.: *Fish Biodiversity of India*, NBFGR, 2002.

Ralph, R.: *Breeding Australia's desert goby*, Tropical Fish Hobbyist, 1988.

Rao, D.V.: *Poisonous and Venomous Fishes of Andaman Islands, Bay of Bengal,* Zoological Survey of India, 2003.

Rebelin, W.E.: *The Pathology of Fishes*, The University of Wisconsin Press, UK, 1975.

Rosen, D. E.: *Modes of Reproduction in Fishes, How Fishes Breed*, Natural History Press, New York, 1966.

Sharma, Arti: *Fishes: Aid to Collection, Preservation and Identification*, Daya, Delhi, 2006.

Singh, Braj Prasad: *Fish Genetic and Biotechnology*, Swastik, Delhi, 2008.

Stoskopf, M.K.: *Fish Medicine*, W.B. Saunders Co. 1993.

Tunstall, Jeremy: *The Fisherman: The Sociology of an Extreme Occupation*, McGibbon and Kee, London, 1969.

Untergasser, D.: *Handbook of Fish Diseases*, T.F.H. Publications, Neptune, NJ, 1989.

VanStone, J. W.: *Athapaskan Adaptations: Hunters and Fishermen of the Subarctic Forests*, Aldine, Chicago, 1974.

Villars, T.: *The Chemical Senses of Fishes*, Academic Press, London, 1974.

Ward, Barbara E.: *Varieties of the Conscious Model: Fishermen of South China,* Tavistock, London, 1965.

Warner, William W.: *Beautiful Swimmers: Watermen, Crabs and the Chesapeake Bay*, Little, Brown, Boston, 1976.

Index

❑❑❑